IMAGING of NUCLEIC ACIDS and QUANTITATION in PHOTONIC MICROSCOPY

Methods in Visualization

Series Editor: Gérard Morel

In Situ Hybridization in Light Microscopy
Gérard Morel and Annie Cavalier

Visualization of Receptors: In Situ Applications of Radioligand Binding
Emmanuel Moyse and Slavica M. Krantic

Genome Visualization by Classic Methods in Light Microscopy
Jean-Marie Exbrayat

Imaging of Nucleic Acids and Quantitation in Photonic Microscopy
Xavier Ronot and Yves Usson

In Situ Hybridization in Electron Microscopy
Gérard Morel, Annie Cavalier, and Lynda Williams

IMAGING of NUCLEIC ACIDS and QUANTITATION in PHOTONIC MICROSCOPY

Edited by
Xavier Ronot
Professor at the Ecole Pratique des Hautes Etudes

Yves Usson
Senior Researcher at the Centre National de la Recherche Scientifique

CRC Press
Boca Raton London New York Washington, D.C.

Cover inset: Confocal 3-D reconstruction of HeLa cell mitotic nuclei after heat shock treatment. chromosomes are shown in red, and HSF1 transcription factors are shown in green. For details, see Jolly, Usson, and Morimoto, PNAS, 1998, 96: 6769–6774.

Library of Congress Cataloging-in-Publication Data

Imaging of nucleic acids and quantitation in photonic microscopy / edited by Yves Usson, Xavier Ronot.
p. cm. — (Methods in visualization)
ISBN 0-8493-0817-8 (alk. paper)
1. Nucleic acids—Analysis. 2. Fluorescent probes. 3. Fluorescence microscopy. I. Usson, Yves. II. Ronot, Xavier. III. Series.

QP620 .I43 2001
572.8'028—dc21 00-140132
CIP

Direct all inquiries to CRC Press LLC, 2000 N.W. Corporate Blvd., Boca Raton, Florida 33431.

International Standard Book Number 0-8493-0817-8
Library of Congress Card Number 00-140132
Printed in the United States of America 1 2 3 4 5 6 7 8 9 0
Printed on acid-free paper

SERIES PREFACE

Visualizing molecules inside organisms, tissues, or cells continues to be an exciting challenge for cell biologists. With new discoveries in physics, chemistry, immunology, pharmacology, molecular biology, analytical methods, etc., limits and possibilities are expanded, not only for older visualizing methods (photonic and electronic microscopy), but also for more recent methods (confocal and scanning tunneling microscopy). These visualization techniques have gained so much in specificity and sensitivity that many researchers are considering expansion from in-tube to *in situ* experiments. The application potentials are expanding not only in pathology applications but also in more-restricted applications such as tridimensional structural analysis or functional genomics.

This series addresses the need for information in this field by presenting theoretical and technical information on a wide variety of related subjects: *in situ* techniques, visualization of structures, localization and interaction of molecules, functional dynamism *in vitro* or *in vivo*.

The tasks involved in developing these methods often deter researchers and students from using them. To overcome this, the techniques are presented with supporting materials such as governing principles, sample preparation, data analysis, and carefully selected protocols. Additionally, at every step we insert guidelines, comments, and pointers on ways to increase sensitivity and specificity, as well as to reduce background noise. Consistent throughout this series is an original two-column presentation with conceptual schematics, synthesizing tables, and useful comments that help the user to quickly locate protocols and identify limits of specific protocols within the parameter being investigated.

The titles in this series are written by experts who provide to both newcomers and seasoned researchers a theoretical and practical approach to cellular biology and empower them with tools to develop or optimize protocols and to visualize their results. The series is useful to the experienced histologist as well as to the student confronting identification or analytical expression problems. It provides technical clues that could only be available through long-time research experience.

Gérard Morel, Ph.D.
Series Editor

PREFACE

The discovery of the cell nucleus in 1831 was the first step in the story of cell and nucleic acid analysis. The cell nucleus governs the biosynthetic events that characterize the cell's past and future. At the molecular level, the cell nucleus is designed to produce, or direct the production of, molecules essential for cell life and necessary for cell proliferation, differentiation, and death. This activity is programmed by key elements: the nucleotide sequences in the deoxyribonucleic acid of the chromatin.

Researchers interested in nuclear structure and function feel the need to establish relationships between morphology and function and activity, and to search for cytological clues to regulatory and control mechanisms. Visualization of nuclear components has introduced new dimensions to the biology of the nucleus and new insights into its chemistry.

Nuclear organization studies follow two main lines: biochemical analysis of isolated nuclei and cytological analysis in living or fixed cells. The second approach probes and explores nuclear components inside the cell using microscopy.

Imaging of Nucleic Acids and Quantitation in Photonic Microscopy was undertaken to serve as a tutorial for all researchers interested in the analysis of nucleic acids by image cytometry following fluorescent labeling and detection procedures.

This book belongs to the series entitled *Methods in Visualization*. Its special feature is to cover theoretical bases and to highlight the critical points, pro and cons, dos and don'ts resulting from the authors' experience. Chapters are written by researchers with extensive experience in the use of photonic microscopy in the field of nucleic acid research.

The design of probes that enable direct or indirect labeling of nucleic acids is discussed in Chapters 1 and 2. Other chapters describe methodologies developed for microscopic analysis of nucleic acids: single cell electrophoresis (comet assay) to detect DNA lesions (Chapter 3), and fluorescent *in situ* hybridization of artificially decondensed DNA (Chapter 4). Chapters 5, 6, and 7 describe fluorescent imaging principles for the *in situ* study of nuclear distribution and colocalization of molecules (epi-illumination, confocal, two-photon microscopes, and associated captors), different steps in image processing in the case of fluorescently labeled objects. Chapter 8 discusses examples of quantification: chromatin organization in cells maintained alive and distribution of nucleolar bodies inside the nucleus (AgNOR).

The amount of information provided by photonic microscopy contributes widely to the understanding of nuclear function and related components.

ACKNOWLEDGMENTS

This book was written in the framework of the European "Leonardo da Vinci" project (Grant F/96/2/0958/PI/II.I.1c/FPC), in association with Claude Bernard-Lyon 1 University (http://brise.ujf-grenoble.fr/LEONARDO).

CONTRIBUTORS

Anne-Sophie Gauchez .... Praticien Hospitalier CHUG — Physician at the Grenoble University Hospital

Didier Grunwald .... Ingénieur CEA — Research officer at the Centre d'Etudes Nucléaires de Grenoble

Karine Monier .... Postdoctoral Fellow — Scripps Research Institute, La Jolla

Amaury du Moulinet d'Hardemare .... Maître de Conférence UJF — Assistant Professor at the Joseph Fourier University — Grenoble 1

Marie-Jeanne Richard .... Praticien Hospitalier CHUG — Physician at the Grenoble University Hospital

Xavier Ronot .... Directeur d'Etudes EPHE — Professor at the Ecole Pratique des Hautes Etudes, Grenoble

Sophie Rousseaux .... Praticien Hospitalier CHUG and Maître de Conférence UJF — Physician at the Grenoble University Hospital and Assistant Professor at the Joseph Fourier University — Grenoble 1

Sylvie Sauvaigo .... Ingénieur CEA — Research officer at the Centre d'Etudes Nucléaires de Grenoble

Yves Usson .... Chargé de recherche CNRS — Research officer at the Centre National de la Recherche Scientifique, Grenoble

CONTENTS

ABBREVIATIONS

2D, 2-D ⇒ two dimensions
2PM ⇒ two-photon microscopy
3D, 3-D ⇒ three dimensions
a.u. ⇒ arbitrary units
A–T ⇒ adenine–thymine
AO ⇒ acridine orange
BrdU ⇒ bromodeoxy uridine
BSA ⇒ bovine serum albumin
CCD ⇒ charge-coupled device
CLSM ⇒ confocal laser scanning microscope
DABCO ⇒ 1,4 diazabicyclo (2.2.2.) octane
DNA ⇒ deoxyribonucleic acid
EB ⇒ ethidium bromide
e^- ⇒ electric charge
FITC ⇒ fluorescein isothiocyanate
FWHM ⇒ full width half maximum
G–C ⇒ guanine–cytosine
HO ⇒ Hoechst
IgG ⇒ immunoglobulin
MW ⇒ molecular weight
NA ⇒ numerical aperture
PBS ⇒ phosphate-buffered saline
PI ⇒ propidium iodide
PM, PMT ⇒ photomultiplier tube
PSF ⇒ point spread function
SNR ⇒ signal vs. noise ratio
RNA ⇒ ribonucleic acid
RNase ⇒ ribonuclease
UV ⇒ ultraviolet

Units

A, mA ⇒ ampere, milliampere
L, mL, µL ⇒ liter, milliliter, microliter
M, m*M*, µ*M* ⇒ molar, millimolar, micromolar
m, mm, µm ⇒ meter, millimeter, micrometer
s, ms, µs, ns, ps, fs ⇒ second, millisecond, microsecond, nanosecond, picosecond, femtosecond
V, mV ⇒ volt, millivolt
W, mW, kW ⇒ watt, milliwatt, kilowatt

Chapter 1

Fluorochromes and Fluorescent Probes

Didier Grunwald
and Xavier Ronot

CONTENTS

1.1 FIXATION

The role of fixation is to preserve the cell and its organelles in a conformation as close as possible to the living state. The main objective is to preserve the structural state of the organelles and their intracellular localization, as well as the proteins–proteins, proteins–nucleic acids interactions, while respecting the necessity imposed by the analytical method.

Fixation is achieved by means of various chemical components. There is no such thing as a universal fixative. Thus, one has to try to choose the fixative best adapted to the experiment goal. It must ensure a good conservation of morphology. Classical fixatives are alcohols (ethanol, methanol), acetone, aldehydes (formaldehyde, paraformaldehyde), acids (acetic, chloride, sulfuric, picric, etc.), metal salts, etc.

⇨ Fixation must allow the fluorochromes to penetrate inside the cells.

1.1.1 Fixatives

1.1.1.1 Ethanol

Ethanol precipitates the proteins (and to a lesser extent, the nucleoproteins), dissolves different lipid complexes, precipitates the glycogen without denaturation, and induces a high cellular contraction.

⇨ In spite of its limitations, the 70% ethanol fixation is a simple and rapid procedure often used before DNA staining.
⇨ It is worth noting that the chondriome and the Golgi apparatus are destroyed, and lipids are solubilized.
⇨ Fixed cells can be stored at 4°C for short periods.
⇨ For longer periods, storage at –20°C is recommended.

1.1.1.2 Formaldehyde

Formaldehyde is a reducing agent which does not precipitate proteins, fixes lipid complexes, and does not contract the structures. In comparison to alcohol, formaldehyde ensures a better conservation of cellular integrity, especially a better fixation of the membranes, with concentration from 1 to 5%. This fixative penetrates rapidly, but the process of fixation is, nevertheless, slower than with ethanol.

⇨ For stoichiometric and rapid staining, a permeabilization step is recommended (e.g., Triton X100: 0.1%).

0-8493-0817-8/01/$0.00+$1.50

1.1.1.3 Glutaraldehyde

Glutaraldehyde fixation is very similar to that obtained with formaldehyde.

⇨ Glutaraldehyde's main disadvantage is introducing autofluorescence.

1.1.1.4 Boehm–Sprenger

Boehm–Sprenger fixative consists of methanol 80%, formaldehyde 15%, and acetic acid 5%.

⇨ Boehm–Sprenger has been described as preserving the chromatin organization.

1.1.1.5 Carnoy

Carnoy, a mixture of methanol 75% and acetic acid 25%, can be an alternative to Boehm–Sprenger.

1.2 FLUOROCHROMES

1.2.1 Principle of Fluorescence

Fluorescence is a complex phenomenon where light is emitted from molecules for a short period of time following the absorption of light. The wavelength of the emitted fluorescence is longer than the wavelength of absorbed light.

⇨ In conventional fluorescence, the emission wavelength is always superior to the excitation wavelength.

⇨ The energy of emitted light is usually weaker than the energy that has been absorbed.

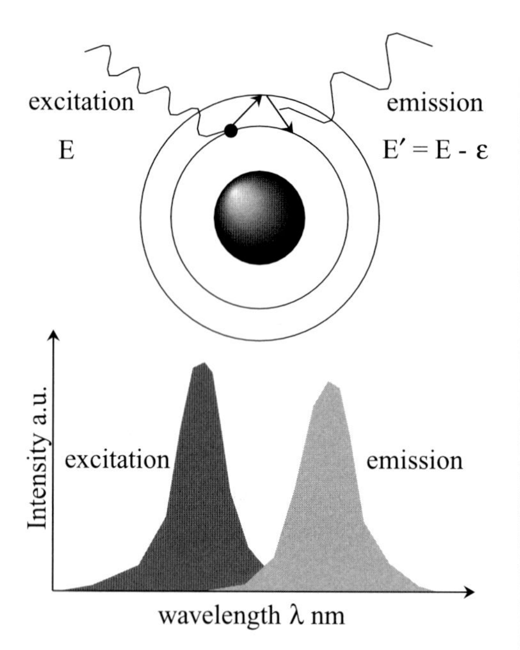

Figure 1.1 Principle of fluorescence.

Excitation spectrum: dark gray
Emission spectrum: light gray

1.2.2 Intercalating Fluorochromes

Among the most frequently used fluorochromes, propidium iodide (PI, 3,8-diamino-5-diethyl-methylaminopropyl-6-phenyl phenantridium diiodide) and ethidium bromide (EB, 2,7-diamino-9-phenyl-10-ethyl phenanthridinium bromide) both intercalate between base pairs of double-stranded DNA and RNA without base specificity. The primary intercalative mode of dye binding depends on double-helical polynucleotide structure. Both can be excited by a visible wavelength (488 nm) and emit a red fluorescence (650 nm) with a high quantum yield.

⇨ For DNA content analysis, labeling requires a prior RNase treatment to eliminate RNA molecules.

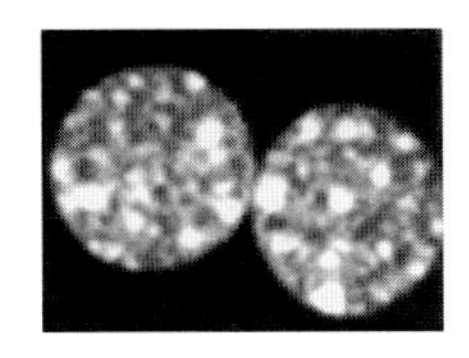

Figure 1.2 DNA staining with propidium iodide.

These dyes are not DNA specific, but also bind to the double-stranded polynucleotide molecules (i.e., DNA–DNA, but also RNA–RNA, and RNA–DNA).

Acridine orange (AO, 3,6-dimethylamino acridine) is a metachromatic dye that emits a green fluorescence (520 nm) when intercalated between DNA base pairs and a red fluorescence (640 nm) when stacked on RNA. Both electrostatic and hydrophobic forces influence its binding to DNA.

⇨ Acridine orange allows the simultaneous measurement of DNA and RNA.

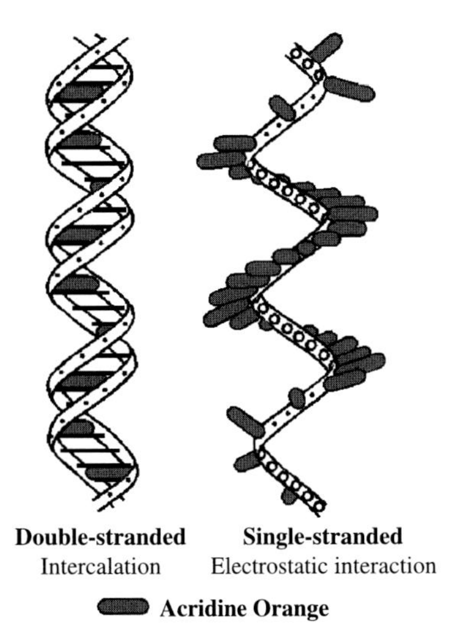

Figure 1.3 Binding mode of acridine orange.

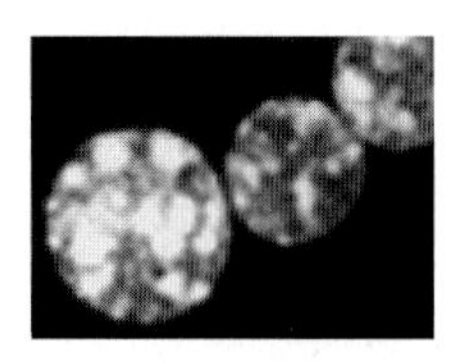

Figure 1.4 DNA staining with acridine orange.

1.2.3 Base-Specific Fluorochromes

⇨ These dyes are strictly DNA specific.

1.2.3.1 Adenine/Thymine (A–T)

DAPI, Hoechst 33258, Hoechst 33342 are benzimidazole derivatives that bind preferentially to A–T base regions. Using a UV light excitation (350 nm), they emit a blue fluorescence, with a maximum near 450 nm.

⇨ Hoechst 33342, considered a "vital" fluorochrome, can be incorporated into the nuclei of living cells.

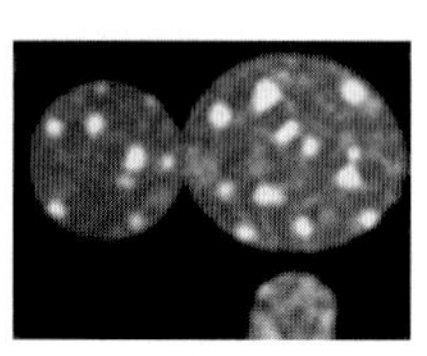

Figure 1.5 DNA staining with Hoechst 33342.

DAPI and Hoechst 33258 cannot stain the DNA of living cells by penetrating through the plasmic membrane. A cell fixation stage is needed prior to nuclear staining.

1.2.3.2 Guanine/Cytosine (G–C)

Mithramycin, chromomycin, and olivomycin display similar binding characteristics, preferentially to G–C base regions by nonintercalating mechanisms.

Mithramycin and chromomycin A3 Mithramycin are green–yellow fluorescent DNA-specific antibiotics.

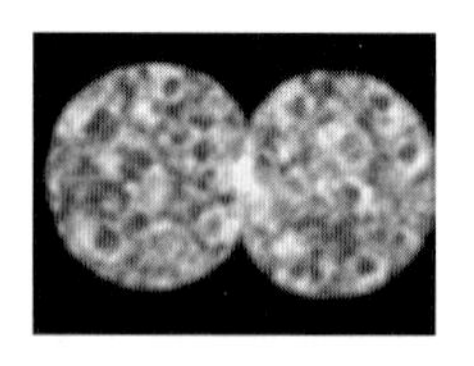

Figure 1.6 DNA staining with mithramycin.

1.2.4 Technical Steps for Propidium Iodide or Ethidium Bromide Staining

1.2.4.1 Material/Products/Protocol

1.2.4.1.1 Material

- Glass slides
- Cytospin
- Fluorescence microscope

1.2.4.1.2 Products

- PBS 1X
- Fixative
- DNA staining

⇨ Use PBS without Ca^{++} and Mg^{++}.
⇨ Ethanol 70% in H_2O
⇨ Propidium iodide or ethidium bromide. Stock solution: 1 mg/mL in H_2O, stored at 4°C. Excitation: 488 nm.

1.2.4.1.3 Protocol

- Cell preparation

⇨ Adherent cells: eliminate culture medium and wash the cell monolayer with PBS.
⇨ Nonadherent cells: centrifuge the cell suspension (5 min, 800 *g*); resuspend the cells in PBS and cytocentrifugate them on a glass slide.

- Fixation — **30 min at 6°C**

⇨ Add ethanol 70% at room temperature.

- DNA staining — **30 min at room temperature**

⇨ Add PBS containing RNase (Sigma, #R4875 Kunitz units, 0.5 µg/mL) and propidium iodide or ethidium bromide.

- Elimination of fluorochrome — **30 min**

⇨ Wash the cells with PBS in order to avoid background fluorescence.

- Mounting

⇨ Mount the slides with a medium containing glycerol 90%, DABCO (Sigma #D2522, 2%, and 0.02% NaN_3).

- Microscopic observation

⇨ Observe the nuclear staining with the appropriate filters set. Avoid long exposure, which induces fluorescence fading.

1.2.5 Technical Steps for Hoechst 33258 or DAPI Staining

1.2.5.1 Material/Products/Solutions

1.2.5.1.1 Material

- Glass slides
- Cytospin
- Fluorescence microscope

1.2.5.1.2 Products

- PBS 1X
- Fixative
- DNA staining

⇨ Use PBS without Ca^{++} and Mg^{++}.
⇨ Ethanol 70% in H_2O
⇨ Use the same protocol as for propidium iodide or ethidium bromide staining, except use no RNase in the staining solution: DAPI or Hoechst 33258 (Sigma, 2 µg/mL). Excitation: 365 nm.

1.2.5.1.3 Solutions

⇨ Mount the slides with a medium containing glycerol 90%, DABCO (Sigma, 2%, and 0.02% NaN_3).

1.3 NUCLEOTIDE SUBSTITUTION

1.3.1 Principle

The detection and measurement of DNA synthesized during the replication phase permit the identification of cells in the S-phase of the cell cycle. This measurement is generally made in association with a measurement of the global DNA content, showing the position of each cell in this phase.

This approach generally uses the thymidine analogue bromodeoxy uridine (BrdU), which possesses antigenic properties. After a short period of growing cells in the presence of BrdU, it is possible to detect the newly synthesized DNA by fluorescent immunostaining.

⇨ Iododeoxy uridine may be used

1.3.2 Technical Steps

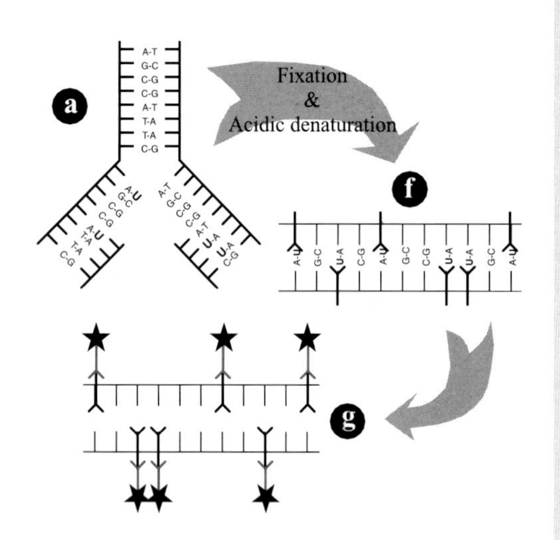

a = **Incubation with BrdU**
b = **Cell fixation**
c = **Rehydration**
d = **Acid denaturation (histone elimination)**
e = **Neutralization**
f = **Labeling with an anti-BrdU antibody**
g = **Detection with a fluorescent secondary antibody**
h = **DNA counterstaining**

Figure 1.7

1.3.3 Material/Products/Solutions

1.3.3.1 Material

- Agitation table
- Centrifuge, centrifugation tubes

⇨ Weak horizontal shaking
⇨ Centrifuge (1000 *g*)

1.3.3.2 Products/Solutions

- PBS 1X — ⇨ Use PBS without Ca^{++} and Mg^{++}.
- PTB — ⇨ PBS, 0.5% Tween 20, 0.5% BSA
- PTBT — ⇨ Sodium tetraborate 0.1 *M* in PBS, pH 8.5, 0.5% Tween 20, 0.5% BSA
- BrdU — ⇨ Stock solution: 10^{-2} *M*, stored at –20°C. MW = 307.1
- Fixative — ⇨ Ethanol 70% in H_2O
- HCl 4N
- Tween 20 — ⇨ Stock solution: 10% in PBS; dilute at time of use
- BSA — ⇨ Stock solution: 5% in PBS + sodium azide 0.02%, pH 7.2; store at 4°C
- Anti-BrdU antibody — ⇨ Becton–Dickinson, #7580
- Anti-IgG/FITC — ⇨ FITC labeled anti-mouse anti-IgG
- DNA staining — ⇨ Hoechst 33258, stock solution: 1 mg/mL in H_2O, stored at 4°C. **Note:** UV excitation
  ⇨ Propidium iodide, stock solution: 1 mg/mL in H_2O, stored at 4°C. Excitation: 488 nm

1.3.4 Protocol

1.3.4.1 BrdU incorporation

- In a cell culture, add BrdU (5 µ*M* final) — **15 min at 37°C**
  ⇨ Increase the incubation time to 30 min if cells develop slowly (>24 hours). **Do not forget that BrdU is toxic (mutagen).**
- Collect cells by centrifugation — **5 min**
  ⇨ 250 *g*
- Fixation — **30 min on ice**
  ⇨ The pellet is resuspended in 200 µL of PBS–glucose (1 g/L); and then, under weak vortex agitation, is added drop by drop to about 5 mL of 70% ethanol at room temperature. If cells are too highly concentrated (>5 × 10^6 cells/mL), add 70% ethanol until concentration is about 10^6 cells/mL. Homogenize, and place on ice for 30 min.
- Rehydration — **3–4 hours at 4°C**
  ⇨ Centrifuge at 1000 *g* for 10 min; resuspend the pellet in PBS–glucose (10^6 cells/mL) and store at 4°C, at least 4 hours for good rehydration.

1.3.4.2 BrdU detection

- Acidic denaturation — **15 min at room temperature**
  ⇨ 2 × 10^6 fixed cells are suspended in 2 mL PBS–glucose.
  ⇨ Add 2 mL HCl 4N.
  ⇨ **Note:** the denaturation time can be modified, depending on the cell type.

- Neutralization **15 min**

  ⇨ Centrifuge at 1000 *g* for 7 min. Resuspend the pellet in 2 mL PTBT.
  ⇨ Centrifuge at 1000 *g* for 7 min. Resuspend the pellet in 2 mL PTB.

- Labeling with the anti-BrdU antibody **30 min at room temperature**

  ⇨ After centrifugation, resuspend the pellet in 200 µL PTB, and add 20 to 40 µL of anti-BrdU antibody.
  ⇨ Incubate under weak agitation.

- Washing of the first antibody **7 min**

  ⇨ Add 4 mL PTB and centrifuge (1000 *g*, 5 min).

- Labeling with the secondary fluorescent antibody **30 min at room temperature**

  ⇨ Resuspend in 200 µL PTB, and add an anti-mouse IgG antibody, FITC conjugated. Final dilution: 1/100 to 1/400

- Washing of the secondary antibody

  ⇨ Add 4 mL PTB and centrifuge (1000 *g*, 5 min).

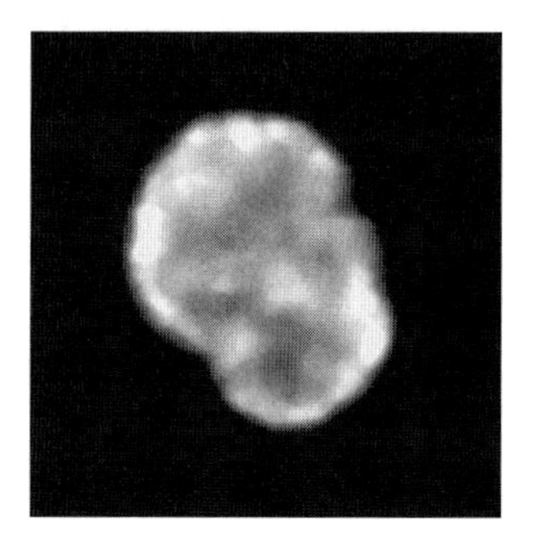

Figure 1.8 Immunofluorescent revelation of BrdU molecules.

- DNA counterstaining **>30 min**

  ⇨ Cells are stained at least 30 min prior to analysis with a DNA-specific fluorochrome (Hoechst 33258 (HO): 2 µg/mL, or propidium iodide (PI): 20 µg/mL). **Note:** HO staining may be done immediately, without any effect on cell conservation; PI staining does not permit cell preservation for longer than 24 hours.

1.4 FLUOROCHROME PROPERTIES

1.4.1 Specificity and Binding Mode

Specificity is the main characteristic of fluorochromes, which allows the detection of a particular cellular component.

⇨ PI or EB fluorochromes are specific for DNA after an RNase treatment, which eliminates RNA molecules

1.4.2 Spectral Properties

Each fluorochrome displays spectral characteristics: for light absorption and fluorescence emission. The absorption spectrum determines optimal wavelength for excitation of the fluorochrome. The choice of filters for fluorescence collection is determined by the emission spectrum of the fluorochrome.

The spectral characteristics of the fluorochromes are modulated by their fixation to the nucleic acid.

⇒ Free and DNA-bound fluorochromes display different spectral characteristics, as well as fluorescence intensity parameters.

Table 1.1 Nucleic Acids Specific Dyes

Name	Wavelength (nm) ↓ excitation max ↑ emission max	Target Nucleic Acid	Base Pair Specificity	Binding Mode
Acridine orange	↓492 - ↑525 (free) ↓500 - ↑524 (bound to DNA) ↓426 - ↑644 (bound to RNA)	DNA and RNA	NONE	Intercalation (double strand) External (single strand)
Ethidium bromide	↓500 - ↑610 (free) ↓510 - ↑595 (bound to DNA)	DNA	NONE	Intercalation
Propidium iodide	↓495 - ↑640 (free) ↓550 - ↑610 (bound to DNA)	DNA	NONE	Intercalation
7-Amino-actinomycin D	↓523 - ↑647 (free) ↓555 - ↑655 (bound to DNA)	DNA	NONE	Intercalation and external
Olivomycin	↓430 - ↑550 (free and bound to DNA)	DNA	G–C	External
Chromomycin A_3	↓425 - ↑580 (free) ↓410 - ↑560 (bound to DNA)	DNA	G–C	External
Mithramycin	↓420 - ↑580 (free) ↓440 - ↑570 (bound to DNA)	DNA	G–C	External
Quinacrine mustard	↓430 - ↑525 (free) ↓430 - ↑500 (bound to DNA)	DNA	G-C	Intercalation and external
DAPI 4′,6-Diamidino-2-phenylindole	↓340 - ↑453 (free) ↓347 - ↑458 (bound to DNA)	DNA	A–T	Intercalation and external
Hoechst 33258	↓338 - ↑505 (free) ↓356 - ↑465 (bound to DNA)	DNA	A–T	External
Hoechst 33342	↓343 - ↑383 (free) ↓340 - ↑450 (bound to DNA)	DNA	A–T	External

Note: A–T: Adeninine–Thymine; G–C: Guanine–Cytosine.

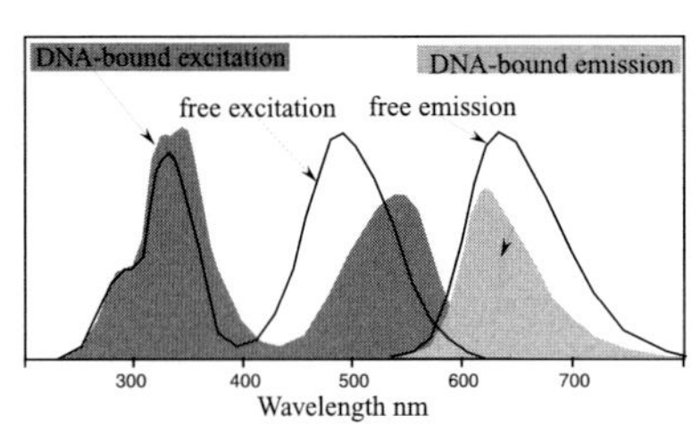

Propidium iodide

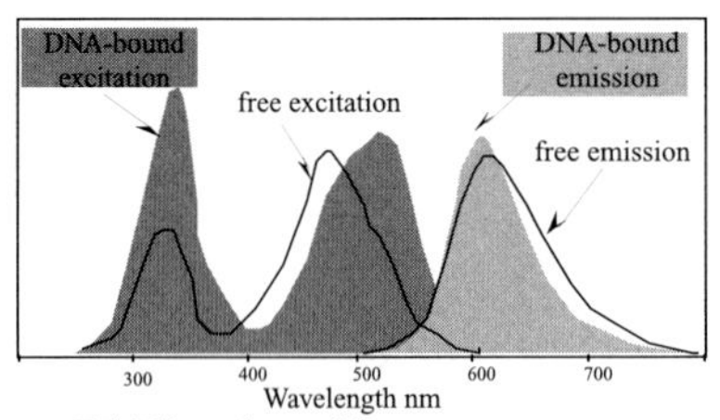

Ethidium bromide

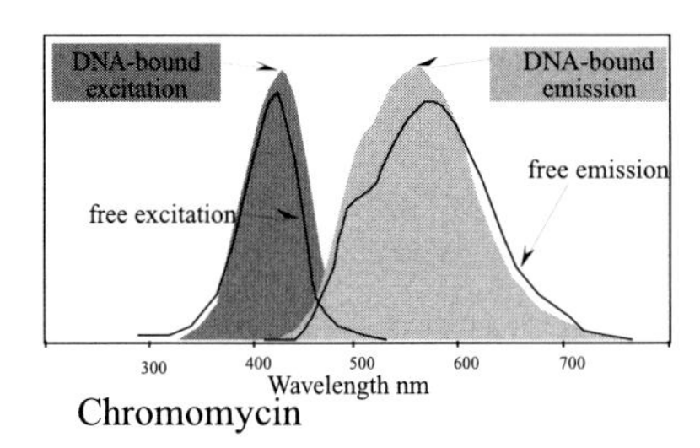

Chromomycin

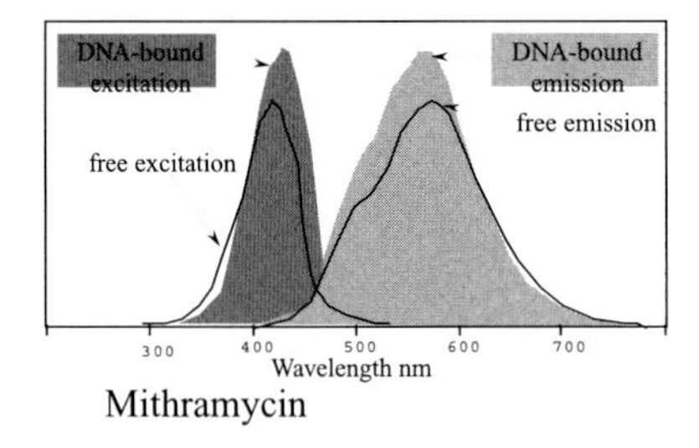

Mithramycin

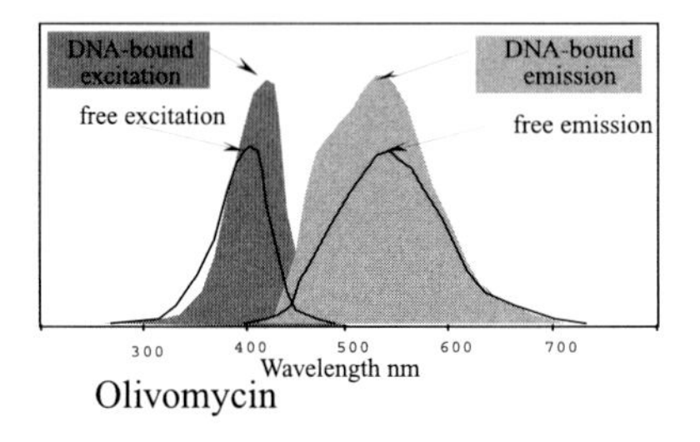

Olivomycin

Figure 1.9 Spectra of various fluorescent probes and fluorophores.

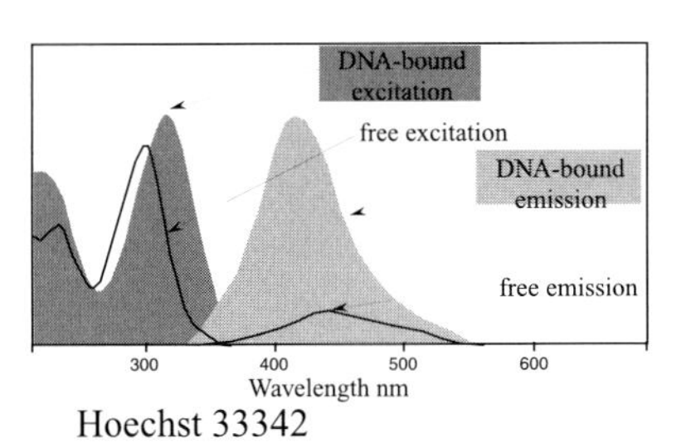

Hoechst 33342

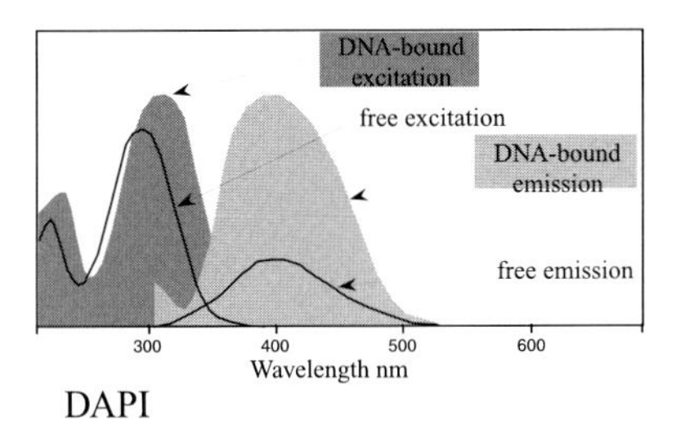

DAPI

Figure 1.9 (continued)

1.4.3 Accessibility

Most fluorochromes specific for DNA can stain this nucleic acid after a cell fixation step. Some fixatives require a permeation step for a stoichiometric labeling.

⇨ Proteins associated with DNA in the nucleus can modify the accessibility of the fixation sites of fluorochromes specific to this nucleic acid.

1.4.4 Stoichiometry

A labeling is stoichiometric when:

- The number of fluorochrome molecules is proportional to the concentration of the nucleic acid of interest
- The measured fluorescence intensity is proportional to the number of fluorochrome molecules
- The fluorochrome binding to the nucleic acid is specific

⇨ The DNA content associated fluorescence of G2+M cells is twice the DNA content associated fluorescence of G0/1 cells.

Stoichiometry can be altered when:

- The labeling procedure is inadequate regarding the fluorochrome concentration, the ionic force, pH, etc.
- The accessibility of fluorochromes to nucleic acid molecules is reduced

⇨ Chromatin organization can modify the accessibility of nucleic acid-specific fluorochromes.

Stoichiometry can be lost when high fluorochrome concentrations are used due to a cooperative effect of fluorochrome molecules, inducing inaccurate measurements of fluorescence.

Light scattering can occur with increasing concentrations of fluorochromes. Use of weak excitation reduces this phenomenon.

1.4.5 Quantum Yield

The quantum yield of a fluorochrome is the ratio of the number of emitted photons to the number of absorbed photons.

The quantum yield is dependent on various parameters such as fluorochrome concentration, pH, ionic concentration, and excitation wavelength.

⇨ A fluorochrome displaying a high quantum yield makes it possible to use a low-level excitation light, thus reducing the fading phenomenon.

1.4.6 Fading and Photobleaching

The principle of fluorescence emission, which implies a de-excitation from the lowest singlet excited state to the ground state, leads to a continuous destruction of fluorescence emission by the incident light.

Experimental conditions are required to limit the fading phenomenon while imaging or quantifying fluorescence.

⇨ A low level of light excitation power contributes to the decrease of the fading phenomenon.

The right choice of mounting medium also contributes to reducing the fading phenomenon, mainly due to the oxidation of the sample. Various anti-fading agents (DABCO, Na_3N, NaI, PPD) in the mounting medium help to eliminate oxygen.

⇨ These antifading agents are not compatible with samples containing living cells.

1.4.7 Toxicity

The toxicity of fluorochromes must be considered with the use of living cells, since they can induce rapid or slow functional perturbations of the cells.

⇨ The effects of fluorochromes on cell viability and proliferation are the widely measured parameters; they are dependent on the fluorochrome concentration, the duration of the labeling, and the retention time.

Chapter 2

Nucleic Probes

Amaury du Moulinet d'Hardemare
and Anne-Sophie Gauchez

CONTENTS

2.1 DEFINITION

Nucleic probes are chemical compounds that can bind to segments of RNA or DNA of varying length. The forces that make such a binding possible are essentially electrostatic, with hydrogen bonds the most important. For example, in a cell, the DNA strands are linked together by hydrogen bonds formed exclusively between the so-called complementary bases, namely, adenine and thymine on the one hand and cytosine and guanine on the other hand (Figure 2.1).

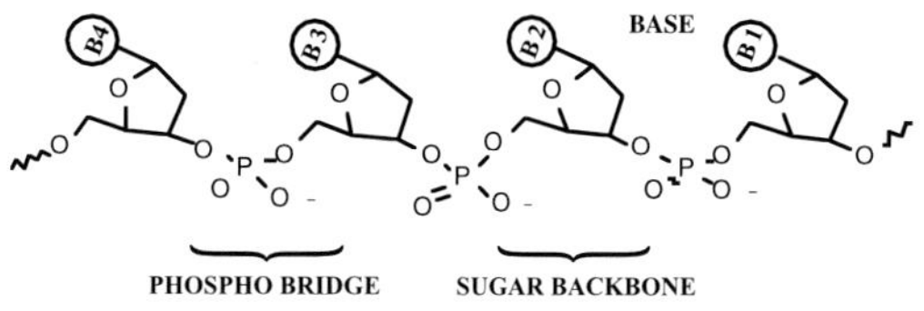

Figure 2.1 Schematic model of natural oligonucleotide.

This interaction is specific and is named after Watson and Crick, who identified this relationship. This structure is the basis of the recognition phenomena used in the development of nucleic probes. Thus, a distinction will be made between:

1. Probes that have only base sequences identical to those of the RNA or DNA strands naturally present in the cell
2. Probes with bases that are modified to varying degrees compared with their natural model, the aim being to modify the physical and chemical properties of the probes.

2.2 NONMODIFIED OLIGONUCLEOTIDES

These are simply DNA strands in which the base sequence is chosen such that it is complementary to that of the target. This target may be either a portion of the DNA of the genome or a portion of the messenger RNA present in the cytoplasm.

0-8493-0817-8/01/$0.00+$1.50

❑ *Advantages*

These oligonucleotides have exactly the same physical, chemical, and biological properties as their target. Oligonucleotides can now be synthesized automatically using commercially available equipment based on solid support chemistry. It is thus possible to obtain relatively large amounts of "natural" oligonucleotides, that is, oligonucleotides that have not been chemically modified.

⇨ Expensive

⇨ Between a milligram and a gram

⇨ For example, they have no fluorescent label.

❑ *Disadvantages*

There are certain disadvantages to their use as biological probes. The most important is their rapid degradation by enzymes which are naturally present in the blood or cytoplasm and which have the role of destroying RNA or DNA that is of no use or even dangerous, such as viral RNA. Once they have been chemically modified, synthesis is much more difficult and consequently not as many probes are produced.

⇨ There are two kinds of such enzymes: exonucleases, which cause cleavage of the oligonucleotides at their 5′ or 3′ ends, and endonucleases, which cleave the oligonucleotides within the sequence itself.

⇨ Reagents are costly and not always readily available.

For applications involving fluorescent probes, oligonucleotides are generally labeled at the 5′ end of the sequence since chemical introduction of the fluorescent label is much simpler in this position. It is possible to introduce these labels using a commercial oligonucleotide through reaction of an activated ester derived from fluorochrome. The ester reacts at the 5′ hydroxyl end, which is less crowded than the 3′ hydroxyl position. In this position, the fluorochrome does not hamper the establishment of the Watson and Crick interactions, and the probe keeps all its recognition capacities.

⇨ In theory, any kind of fluorescent group can be used, but in practice only derivatives of fluoroscein, rhodamines, and a few cyanines are used.

⇨ But still remains subject to enzymatic degradation through the endo- and exonucleases

2.3 CHEMICALLY MODIFIED NUCLEOTIDES

2.3.1 Principle

In order to improve the characteristics of the probes, oligonucleotides can be modified to varying degrees. Such modifications are designed to enhance stability with respect to nucleases and, to a lesser extent, increase the intensity of the Watson and Crick interactions (thus improving the stability of the probe/target double strand). To this end, a great many modifications have been made to oligonucleotides. They are too numerous to describe in detail here, but they are documented in several excellent reviews. The modifications made to oligonucleotides can be divided into three groups, which in fact correspond to the three chemical components of natural oligonucleotides: the phosphate bridge, the bases, and the sugar backbone.

⇨ *See* Figure 2.1.

2.3.2 Modification of the Phosphate Bridge

As we have already mentioned, this type of modification is often made to inhibit or eliminate *in vivo* cleavage of the phosphodiester linkage by the nucleases, which is the reason for the rapid degradation of the oligonucleotides. But such modifications are also made to facilitate uptake by the cells. Cellular uptake can be improved by eliminating the negative charge of the phosphates, and the oligonucleotides thus formed can diffuse passively across the membranes. Two types of modification may be made:

1. Partial modification of the phosphate bridge or total modification to eliminate the phosphorus atoms (dephospho bridges).
2. Partial modifications of the phosphate bridge concern phosphotriesters, methylphosphonates, phosphorothioates, and phosphoramidates.

⇨ *See* Figure 2.2

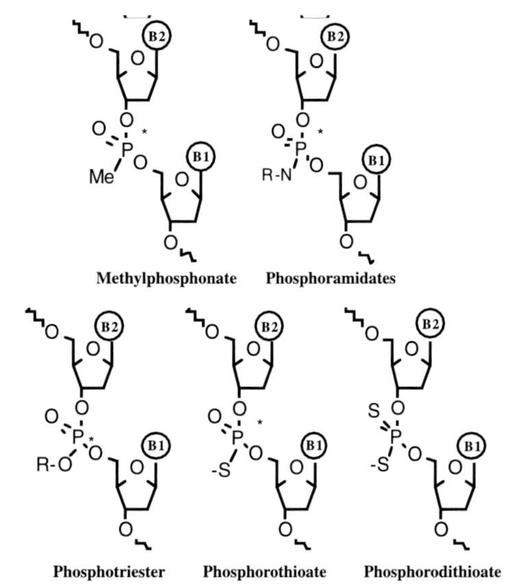

Figure 2.2 Principal modifications to phosphate bridge.

With such modifications, an increase in resistance to nucleases and improved cellular uptake are obtained. While in the case of the phosphotriesters, methylphosphonates, and phosphoramidates, the improved uptake is due to the disappearance of the anionic charges, the anionic charge is maintained in the case of the phosphorothioates (mono or dithioates). However, the presence of sulfur causes an interaction with the transmembrane proteins that would then promote endocytosis. One of the disadvantages of this type of modification is the introduction of phosphorus stereocenters (marked with an asterisk in Figure 2.2), which alter molecular recognition and reduce the melting temperature and the stability of the duplex formed with the target RNA, and, in the case of thiophosphates, cause nonspecific interactions with proteins.

Dephospho bridges

The presence of the phosphate bridge is not essential for recognition since duplexes can still be formed even if it is removed. These are so-called "second-generation" modifications. The chemical nature of these dephosphorylated internucleotidic links is varied. For example, some workers propose dimethylene-sulfide, -sulfoxide, or sulfone analogues as a phosphodiester linkage replacement group.

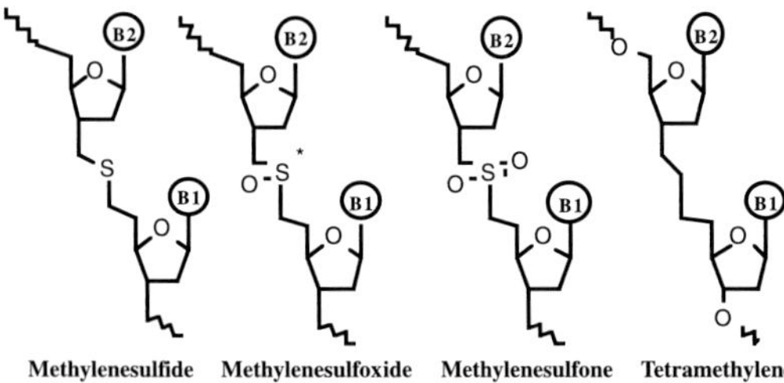

Figure 2.3 Dephospho bridges.

Thanks to these functional groups, it is possible to prepare isoteric, nonionic oligonucleotide analogues that also have remarkable enzymatic resistance and improved chemical stability. Descriptions of other similar dephosphorylated internucleotidic linkages can be found in the literature.

⇨ Siloloxane, formacetal, carbonate, carboxymethyl, carbamate, acetamidate, methylene, etc.

As before, these modifications ensure good resistance to nucleases but at the expense of a reduction in coupling capability with RNA (with the exception of the thioformacetal and methylhydroxylamine links). This observation can be explained in the case of amide and carbamate links by the excessive increase in the stiffness of the backbone obtained by additional conjugation of the N, O hetero-atom of the amide and carbamate function.

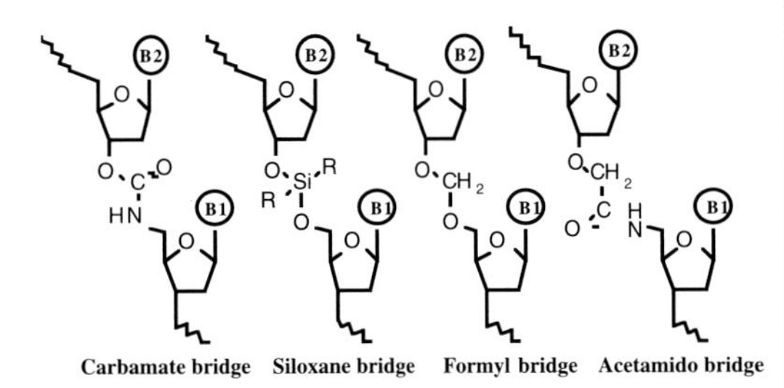

Figure 2.4 Other dephospho bridges.

2.3.3 Modification of Bases

Numerous modified bases have been prepared in order to study the hybridization capability of the derived oligonucleotides or to obtain new pharmacological properties (therapeutic optics). In antisense applications, it is not really a good idea to modify the chemical bases, since such modifications reduce the coupling capability of the complementary bases demonstrated in Watson and Crick's model, undoubtedly resulting in decreased recognition of the target sequence and a lower melting temperature (Tm). However, there are a number of examples where the opposite has been observed, namely, an increase in the stability of the duplexes when modified bases were introduced. Other authors have also observed improved oligonucleotide–RNA hybridization with the substitution of adenosine with 2-aminoadenine.

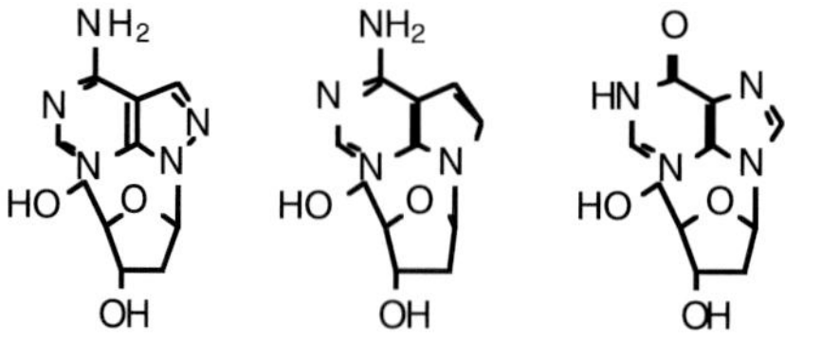

8-aza-7-deaza-2′-desoxyadenine 2′-desoxytubercidine 2′-desoxyinosine

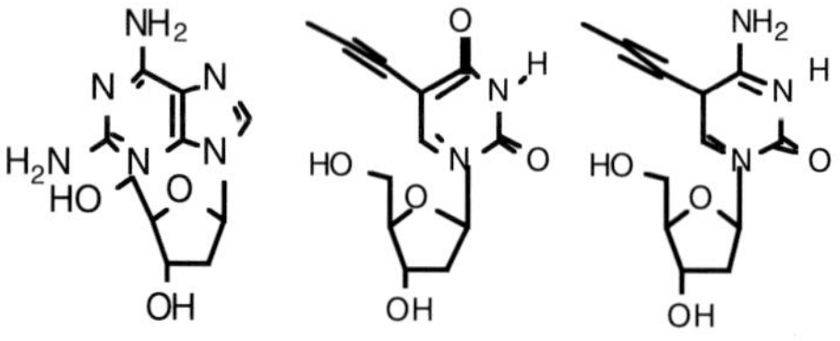

2-amino-desoxyadenosine Acetylenic derivative of bases

Figure 2.5 Some unnatural bases.

Modified base synthesis has also been conducted to facilitate labeling with a radio-element such as iodine. In this case, the base is used as a medium for the label but it can only be introduced at the 3′ or 5′ position; otherwise the oligonucleotide will lose its capacity to form a stable duplex. If the sugar–base combination is conserved, recognition of the phosphorylation enzyme of the 5′ end is possible.

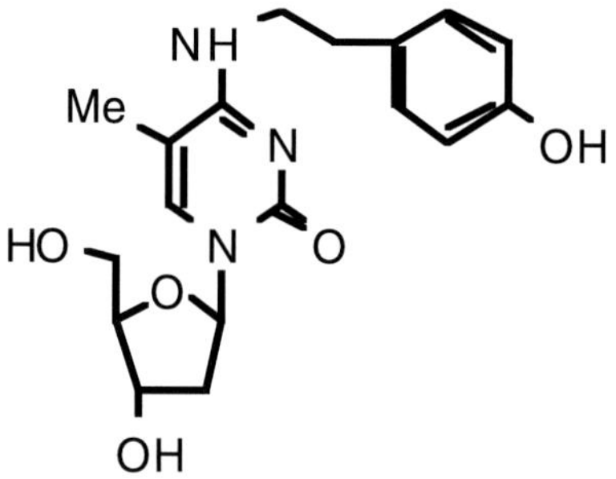

➪ For labeling with radioactive iodine: electrophilic iodine is fixed on the aromatic group by a covalent bond. Electrophilic iodine is generated by oxidation of I^- with chloramine-T, usually in water.

Figure 2.6 Base modified with a tyrosinyl residue.

2.3.4 Modification of the Ribose Sugar Backbone

As in the case of the phosphate bridge, modifications which affect sugars can vary in their extent, a fact which underlines the minor significance of the chemical nature of the ribose sugar backbone in hybridization. Only the isostere aspect and the stiffness of the spacer, important for the correct positioning of the spacer, would appear to be important. It is thus possible to use

2′-*O*-alkylated nucleosides, 2′-alkylated nucleosides, α-anomers of natural nucleosides, or even non-osidic carbocyclic compounds. As in the previous cases, these modifications improve resistance to nucleases but, while an increase in Tm has been observed for the 2′-*O*-alkylated derivatives, the Tm is decreased for the 2′-alkylated derivatives.

⇒ *See* Figure 2.7.

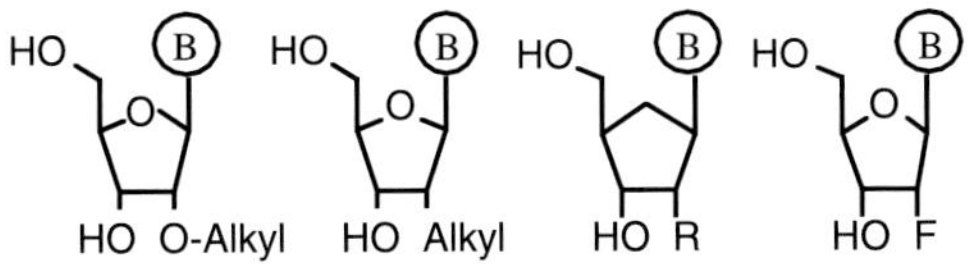

HO O-Alkyl HO Alkyl HO R HO F

Figure 2.7 Ribose derivatives. From left to right: 2′-*O*-alkyl-, 2′-alkyldesoxy-, carboribose, and 2′-fluoro-ribose.

The ribose sugar backbone can also be replaced by the morpholino group without preventing hybridization. Thus, since the presence of the phosphate linkage or ribose sugar backbone is not essential for good hybridization, it would appear that it can be done without both of them. In fact, this is the case with peptide nucleic acids (PNAs), where the bases are linked by an amino acid! Although this is a major modification from a chemical point of view, it seems to be the most promising at the moment. In fact, this modification increases chemical and enzymatic resistance as well as cellular penetration capacity (neutral oligonucleotides), but unlike many of the modifications described previously, it does not introduce a stereocenter or reduce the melting temperature.

⇒ There will be a reduction in Tm.

⇒ *See* Figure 2.8.

⇒ *See* Figure 2.8.

⇒ In fact, the Tm of the PNAs–RNA duplexes or triplexes formed is 10 to 20°C higher than in the case of natural DNA–RNA duplexes.

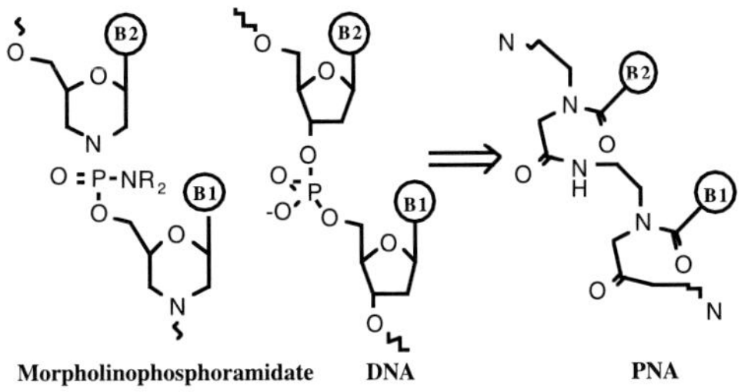

Morpholinophosphoramidate DNA PNA

Figure 2.8 Structural analogy between morpholino DNA, natural DNA, and PNA.

It is also possible to make modifications on an individual basis, in particular at the 3′ and 5′ ends of the oligonucleotide chain where an osidic hydroxyl group is used to bind the markers (anchorage function of linker). Here we are dealing with oligonucleotide conjugates. Conjugation is generally performed by means of a linker with a reactive chemical function, such as *N*-hydroxysuccinimide activated esters or tetrafluorophenol esters. Thus, the molecular probes obtained can be localized by labeling them with a fluorescent label, for example, fluorescein. But the oligonucleotide can also be conjugated with molecules capable of reacting with the target, as in the case of alkylating groups or photoreactive molecules, such as psoralen.

2.4 METHODS OF INCORPORATION

2.4.1 Introduction

Undoubtedly the biggest problem yet to be solved concerning the use of oligonucleotides relates to their penetration into the cells. The probes are in fact polycharged at physiologic pH (phosphate charges and sometimes ammonium base charges). This inhibits their passive diffusion across the cell membrane, which is by nature of low charge and lipophilic. Moreover, there are no channels or natural proteins that can actively or passively transport the oligonucleotides. To date, there is no universal method for ensuring penetration of probes into the cell, although a number of techniques have been developed to facilitate their entry. These techniques can be distinguished according to the level of cell penetration observed, in other words, according to the barrier the probe must cross, which is initially the cell membrane, then the membrane of the cell nucleus. If, in the first instance, visualization of the cytoplasmic targets is desirable (messenger RNAs, for example), in the second case it is the genome itself which is the target to be detected, and thus both membranes have to be crossed.

2.4.2 Crossing the Cell Membrane

2.4.2.1 Viral vector

Bacteriophages or phages are bacterial viruses. They have the advantage of multiplying very quickly, resulting in a considerable number of copies per cell. In addition, they have a system whereby they can enter the bacteria where they

⇨ The bacteria are cultivated up to a density of 4×10^8 mL^{-1} (1 unit of DO600), separated by centrifugation and placed in suspension in a new culture medium containing $MgCl_2$ (10 m*M*). The phage to be amplified is added

multiply autonomously. The yield from this method of infection is significantly higher than in the case of transformation of bacteria by plasmids, and it is possible to insert large DNA fragments. Ultimately, the proliferation of the phage in the bacteria results in lysis of the bacteria, and thus we observe areas of lysis on cultures in gelose.

on the basis of 5×10^7 phages/mL of bacteria. Following adsorption of the phage by the bacteria for 5 min at 37°C, a large volume of culture is obtained. The bacteria are then infected by the phages that multiply actively until lysis of the host bacteria is produced. After addition of CCl_4 and NaCl, the bacterial debris are removed by centrifugation. The phages are precipitated with PEG 600, purified by ultracentrifugation in a $CsCl_2$ gradient. The phages then migrate in a ring 2 to 3 mm thick that is recovered by aspiration through the wall of the centrifugation tube. Between 10^9 and 10^{10} phages/mL of culture medium are obtained in this manner.

⇨ **Phages in $CsCl_2$ at +4°C can be kept for many years.**

⇨ Next, the capsid is destroyed in the presence of proteinase K and SDS. The DNA is phenol-extracted and precipitated in alcohol.

2.4.2.2 Chemical agents of endocytosis or fusion

Since the oligonucleotides have very little interaction with the cell membrane, they are unlikely to cross it. Chemical agents of endocytosis enable the oligonucleotides to interact with the cell membrane, leading to the formation of an endocytosis vacuole which, once integrated in the cytoplasm, liberates the probe. Polycations are water-soluble polymers, such as polyethylenimine, charged at physiologic pH. The cationic chain interacts (coulomb force) with the anionic charges of the phosphodiesters of the probe. Overall, the polymer contributes more positive charges than required for neutralizing the phosphates, and the overall result appears positive. Thanks to this charge, the conjugate is attracted by the transmembrane potential, which is negative in the cell.

Once it has been adsorbed at the surface, the molecular complex triggers an integration phenomenon through a vacuolar endocytosis mechanism. Once inside the cell, the vacuole discharges its contents in the cytoplasm, still under the influence of the polycation charges. The oligonucleotide is thus released into the cytoplasmic compartment.

⇨ Lipophilic polycations and cations, liposomes, lipophilic vectors

It is also possible to guide the cellular binding of probes by conjugating them with a very lipophilic molecule such as a cholesterol molecule. The cholesterol part is adsorbed on the cell membrane, which is itself lipophilic. It is important to note that the oligonucleotide would first bind to the outside of the cell, with the possibility of entering the cell during a second stage. From this point of view, liposome formation is more satisfactory. Liposomes are spherical particles formed by microscopic lipid bilayers in which the probes or other products are inserted. Liposomes have the special capability of being able to fuse with the cell membrane and so release their contents into the cell.

2.4.3 Integration in the Genome

Plasmids are small extra-chromosomal circular fragments of DNA of bacterial origin. They replicate independently of the chromosomes and can be transferred by conjugation. They are between 2 and 5 kilobases (kb) in size, so that they contain only a very small number of genes. These genes can give the bacteria additional properties, for example, resistance to an antibiotic.

A given bacterium can simultaneously contain different plasmids, without endangering its own survival. This type of vector can receive up to 8 or 9 kb of exogenous DNA. In general, the smaller the plasmid, the more rapidly it will replicate and the larger the quantity that can be contained in the bacterium.

⇨ The bacteria containing plasmids are cultivated in a large volume (500 mL to 2 L). Once 2.5×10^8 bacteria/mL (0.6 DO600) have been obtained, an antibiotic (chloramphenicol) is added to inhibit bacterial growth without affecting the replication of the plasmid DNA. The bacterial membrane is then made permeable by various means: lysozyme, streptolysin, electroporation, etc.

⇨ *Preparation of plasmid for recombination:* The plasmid is opened thanks to a restriction enzyme with an individual site. One end of the plasmid is dephosphorylated and the phosphorylated foreign DNA is inserted. The plasmid DNA containing the foreign DNA is closed thanks to a ligase. The DNA to be inserted is of low concentration compared with the plasmid in order to promote combination of foreign DNA and plasmid DNA.

⇨ The bacteria are cultivated in a liquid medium and made permeable to the foreign DNA at 0°C in the presence of $CaCl_2$ at 50 m*M*.

⇨ These transformed bacteria are able to incorporate the foreign DNA and are now competent bacteria. The yield of this transformation is low (1 bacterium for 10^5 to 10^7 μg^{-1} of plasmid). This selection is achieved through cultivation in a selective medium by using the property or properties contributed by the plasmid.

⇒ In general, this is resistance to an antibiotic. The bacteria that have incorporated the recombinant plasmid then become resistant to this antibiotic thanks to the resistance gene of the plasmid. This property is thus proof that the plasmid has been incorporated. In practice, two selection markers are used.

⇒ *Two selection markers:* the foreign DNA is inserted into the 2nd selection marker. When this happens, the gene is activated.

- **Antibiotic marker:** the bacteria that have integrated a recombinant plasmid containing an inserted DNA will be resistant to the first antibiotic but sensitive to the second. On the other hand, bacteria that have incorporated a plasmid with no inserted DNA will be sensitive to both antibiotics. The most classic example is the system combining ampicillin and tetracycline.
- **Lactose operon:** transformation of the *lac* bacterium by a plasmid containing the *lac Z* gene (β galactosidase) gives the bacterium the phenotype *lac+,* which can be characterized directly by means of the bacterial clones thanks to a chromogen substrate.

⇒ The bacteria are cultivated in the presence of IPTG (nonmetabolizable inducer of lactose operon) and X-gal, a galactoside which changes from colorless to blue when it is cleaved by β-galactosidase. In such a medium, the *lac+* bacteria will be blue since they metabolize the X-gal. On the other hand, *lac-* bacteria will be whitish in color. Insertion of a foreign DNA into the *lac Z* inactivates it, and the corresponding colonies remain white.

Chapter 3

Comet Assay

Sylvie Sauvaigo and
Marie-Jeanne Richard

CONTENTS

3.1 INTRODUCTION

The maintenance of DNA integrity is essential for cellular life. Consequently, different approaches are used to check it.

Recent developments are based on separation techniques coupled with very sensitive detection systems (high-performance liquid chromatography, mass spectrometry, electrochemistry). They allow the detection and quantification of different DNA lesions.

Nevertheless, these latter techniques are not adapted to the detection of DNA damage at the cellular level. Moreover, they require a large quantity of cell or tissue material.

As an alternative, the comet assay represents an interesting approach for the study of the genome integrity at the cellular level.

⇨ Quantitative studies of the integrity of cellular DNA are often impaired by two factors: the extraction of the nuclear DNA and the quantity of DNA material necessary for the analysis.

3.2 PRINCIPLE

The comet assay, also known as the single cell gel electrophoresis (SCGE) assay, is a simple and sensitive technique for measuring alkali-labile sites and strand breaks in individual cells.

The cells are embedded in agarose gel on a microscope slide. A lysis solution removes the cell contents except for nuclear material. An electric current is then applied. Damaged DNA migrates out from the nuclei in the direction of the anode and gives the cellular DNA the appearance of a comet (Figure 3.1).

⇨ The earliest version aimed at the detection of DNA double-strand breaks.

⇨ The introduction of alkaline conditions for lysis and electrophoresis extended the scope of the technique to the detection of single strand breaks and alkali-labile sites.

⇨ Applications in the cellular response field (repair studies and apoptosis induction) are also emerging.

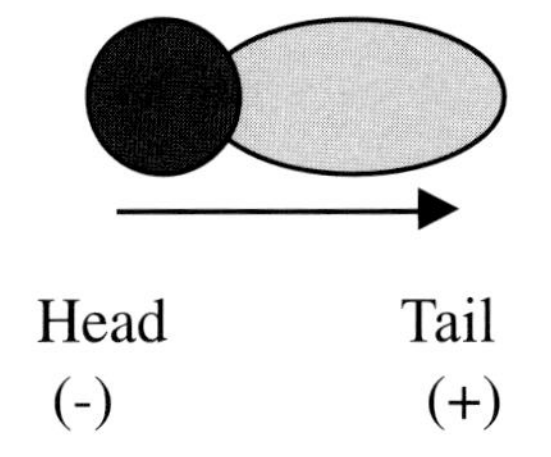

Figure 3.1 Migration of fragmented DNA.

0-8493-0817-8/01/$0.00+$1.50

Because each cell is scored independently, this technique can detect the heterogeneity in the response to a given stress in a cell population.

⇰ Different variants (coupling with lesion-specific antibody detection, DNA digestion with lesion-specific repair enzymes) improve the specificity and the range of this assay.

3.3 COMET ASSAY STEP BY STEP

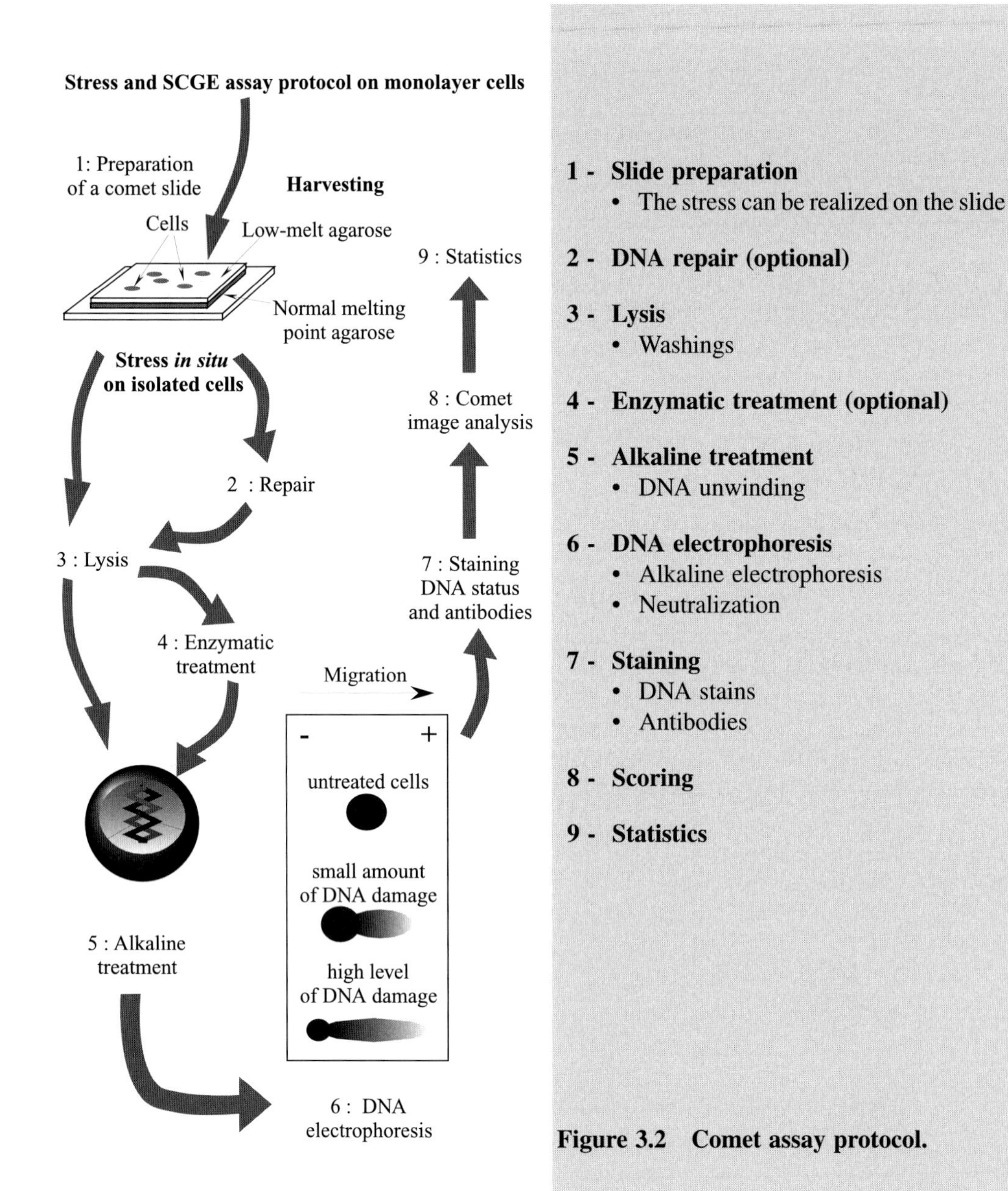

Figure 3.2 Comet assay protocol.

3.4 EQUIPMENT/PRODUCTS/SOLUTIONS

3.4.1 Equipment

- Fully frosted slides and 24 × 50 mm coverslips
  ⇨ Wash the slides with alcohol before use.
- Water baths: 37° and 50°C
- Gel electrophoresis tank
  ⇨ Horizontal tank 20 × 30 cm
- Power pack
  ⇨ Capable of delivering 300 mA at 25 V
- Fluorescence microscope
  ⇨ Equipped with 10× and 40× objectives
- CCD camera
  ⇨ Sensitive system
- Image analysis system
  ⇨ Allows an objective analysis
- Refrigerator
  ⇨ For the lysis step
- Plastic boxes
  ⇨ To hold the slides, protected from light

3.4.2 Products

⇨ High-quality components only

- EDTA-Na_2
- HCl
- Agarose
- Low-melting-point agarose
- NaCl
- DMSO
- PBS without Ca^{2+} and Mg^{2+}
- TRIS acid
- Triton X-100
- Sodium sarcosinate
- NaOH

3.4.3 Solutions

- Lysis solution
  - NaCl 2.5 *M*
  - EDTA 100 m*M*
  - TRIS 10 m*M*
  - Sodium sarcosinate 1%
  - Triton X-100 1%
  - DMSO 10%

⇨ Make up NaCl, EDTA, and TRIS in 800 mL of distilled H_2O. Adjust pH to 10 using approximately 12 g of NaOH. Add the sodium sarcosinate. Adjust to 890 mL. Store at room temperature. Add the Triton X-100 and the DMSO to the desired volume, just before use.

- Electrophoresis buffer
  - NaOH 300 m*M*
  - EDTA-Na_2 1 m*M*

⇨ It is possible to prepare NaOH and EDTA-Na_2 pH10 stock solution (store at +4°C) and to make up the final dilution when required.

- Neutralization buffer
  - TRIS 0.4 *M*, pH 7.5

⇨ Store at +4°C.

- Phosphate buffer (PBS) without Ca^{2+} and Mg^{2+}

⇨ Store at +4°C. Use at room temperature.

- Normal agarose 1% (w/v) in PBS

⇨ After dissolving, maintain this solution in a water bath at 50°C.

- Low-melting-point agarose in PBS

⇨ Maintain this solution in a water bath at 37°C. The agarose concentration can vary (0.6 to 1.2% w/v), depending on the application and on the cell type. The comet tail length will depend on agarose concentration.

3.5 STANDARD PROTOCOL: STRAND-BREAK AND ALKALI-LABILE SITES DETECTION

3.5.1 Cell Preparation

3.5.1.1 Aim

The cells are embedded in agarose:

- To allow their separation
- To preserve their viability

⇨ The first normal agarose layer improves the adhesion of the low-melting-point agarose containing the cells.

3.5.1.2 Step 1: First agarose layer

- Arrange the slides horizontally at room temperature.
- Identify the slides.

⇨ The slides can be prewarmed (40 to 50°C) to facilitate spreading the agarose.

- Add 150 μL of the agarose 1% to the slide.

⇨ This volume can be adapted.

- Immediately lay a coverslip over the agarose.

⇨ Be careful to avoid the formation of bubbles and control the uniformity of the agarose layer.

- Place the slides on ice for 10 min.

⇨ It is possible to store the slides with the first agarose layer at +4°C in a humidified chamber.

3.5.1.3 Step 2: Cell preparation

- Make a cell suspension at a density of 200,000 cells /mL in PBS.

⇨ Cellular medium may be used instead of PBS.

- For each slide, just before use, mix 75 μL of the cell suspension with the same volume of low-melting-point agarose 2X. Gently homogenize the suspension.

⇨ 10,000 to 20,000 cells per slide are necessary for the analysis.

3.5.1.4 Step 3: Cell inclusion

- Take off the coverslip.
- Put 110 µL of the cell suspension on the first agarose layer.
- Cover with a coverslip.
- Leave on ice protected from light.

⇨ The cells may be included between two agarose layers. In this latter case:
- Put 75 µL of the cell suspension in low-melting-point agarose (10,000 to 20,000 cells) on the first agarose layer.
- Cover with a coverslip.
- Place on ice protected from light.
- Gently slide off the coverslip.

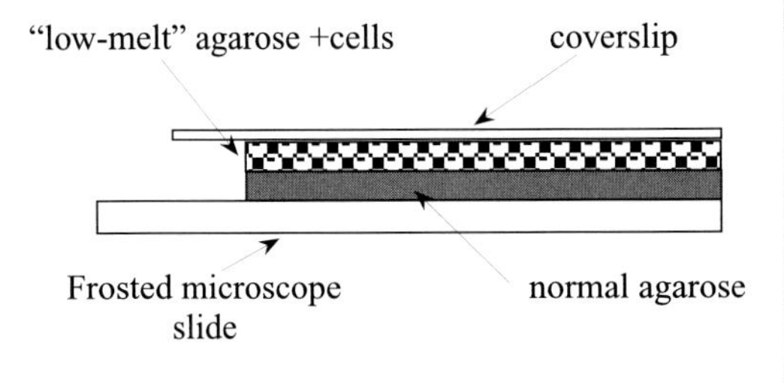

Figure 3.3 Cell inclusion.

⇨ Put 75 µL of low-melting-point agarose
⇨ Replace a coverslip.

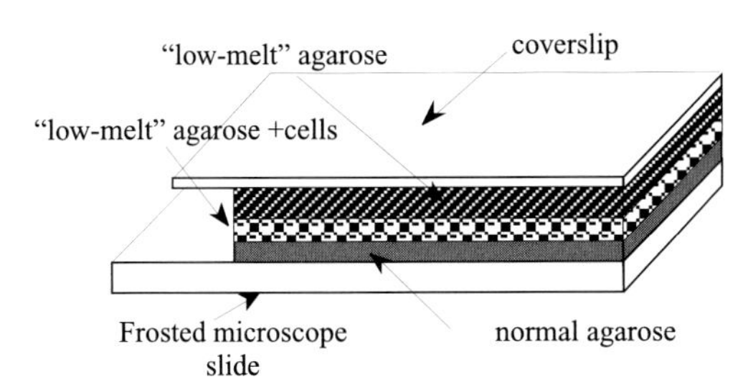

Figure 3.4 Embedded cells.

3.5.2 Membrane Lysis

3.5.2.1 Aim

The nuclear and plasmatic membranes must be eliminated in order to give access to the chromatin.

3.5.2.2 Protocol

- Just before use, add the DMSO and the Triton X-100 to the cold lysis buffer.
- Take off the coverslip.
- Place the slides horizontally in plastic boxes.
- Cover the slides with the lysis solution.
- Store at +4°C for at least 1 hour.
- Slides must be protected from light.

⇨ Mix well; a precipitate may form but it does not alter the properties of the lysis buffer.
⇨ Be careful not to alter the gel.
⇨ Do not put in a vertical position.

⇨ The lysis step can last overnight.
⇨ All the previous steps must be performed under reduced light.

3.5.3 DNA Electrophoresis

3.5.3.1 Aim

During this step the broken DNA fragments are separated through the agarose gel.

According to the standard protocol, the electrophoresis is performed in alkaline buffer. The applied electrophoresis conditions must be standardized to ensure reproducibility between experiments.

⇨ **Option:** electrophoresis may be performed in neutral buffer for the detection of double strand breaks.

3.5.3.2 Electrophoresis protocol

- Gently take off the lysis buffer; aspirate if necessary.
- Place the slides horizontally in the electrophoresis tank.
- Position the slides close together and as close as possible to the anode.
- Gently fill the electrophoresis tank with the buffer; just cover the slides.
- Set for 40 min in the buffer.
- Apply a 25 V tension and a current of 300 mA for 30 min.

⇨ Do not contaminate the electrophoresis buffer with the lysis buffer.

⇨ DNA migration must be performed in the direction of the length of the slides. Identify the migration direction on the slides.

⇨ Fill the empty spaces with blank slides in order to ensure homogenous current and homogenous DNA migration.

⇨ The time of the electrophoresis step must be optimized (20 to 60 min) depending on the cells and the application.

⇨ Adjust the current and the voltage with the buffer level.

3.5.3.3 Neutralization protocol

- Carefully remove the slides from the tank.
- Put them on a tray.
- Cover with neutralization buffer for 5 min.
- Repeat twice.

⇨ Check the pH at the end of the neutralization step with pH paper.

3.5.4 Staining

3.5.4.1 Aim

The objective is to visualize intact and fragmented DNA that has migrated through the gel. Standard techniques use specific DNA dyes.

An option of the assay uses immunological revelation to detect specific DNA lesions.

⇨ Damaged DNA gives the cell the appearance of a comet, with intact DNA in the head.

⇨ This latter method is limited by the availability of specific antibodies. A first application of this variation described the detection of DNA photoproducts (cyclobutane pyrimidine dimers).

3.5.4.2 Products

- Ethidium bromide 20 μg/mL

⇒ This product is highly toxic.
⇒ A stock solution (10 mg/mL) may be stored protected from light.

- DAPI (4,6-diamidino-2-phenylindole) 5 μg/mL
- Hoechst 33258 10 μg/mL

⇒ Store the stock solution at +4°C protected from light.
⇒ Stock solutions of the DNA dyes must be diluted in PBS.

- Lesion-specific primary antibody

⇒ The dilution must be performed just before use (the dilutions used for immunohistochemistry protocols are usually suitable).

- Fluorochrome-labeled secondary antibody

⇒ The fluorochrome choice depends on the available microscope filters (for example, Cy3).

- Washing buffer: PBS containing 0.05% Tween 20

3.5.4.3 Staining protocols

1. DNA dyes
   a. Remove excess buffer.
   b. Add 50 μL of the ethidium bromide solution (or of the dye of choice).

   ⇒ The dye must be stable during excitation.

   c. Cover with a coverslip.
   d. Store the slides in a humidified atmosphere until scoring.

   ⇒ The slides may be stored for 48 to 72 h. It is possible to keep the slides for a longer time after dehydration in an ethanol bath and rehydration before scoring.

2. Specific antibodies
   a. The pH value must be corrected to fit with the standard immunological protocols.
      - Remove excess neutralization buffer.

      ⇒ Excess buffer could lead to the dilution of the antibody.

      - Equilibrate the slides in PBS (3 × 5 min).
      - Aspirate excess PBS.
      - Let the slides air dry for 5 min.

      ⇒ Keep the slides protected from light.

   b. Primary antibody
      - Put 80 μL of the diluted antibody on each slide.
      - Place a coverslip.
      - Incubate for 2 h at 37°C in a closed, humidified box.

      ⇒ The slides may be placed over aluminum foil covering a moist tissue to prevent them from drying.

   c. Secondary antibody
      - Place the slides on ice for 5 min. Take off the coverslips and put them on a tray.

- Wash 3 times for 5 min each with the PBS/Tween 0.05% buffer.
- Remove excess buffer and allow to air dry for 5 min.
- Put 80 µL of secondary antibody dilution on each slide.
- Replace a coverslip.
- Incubate for 2 h at 37°C in a closed, plastic box.
- Set on ice for 5 min.
- Wash 3 times with PBS/Tween 0.05% (5 min each).

d. DNA localization

The emission wavelength of the DNA dye and of the secondary antibody fluorochrome must be different.

⇨ **Optional**

⇨ Maintain in a humidified atmosphere.

⇨ Example, Hoechst 33258 10 µg/mL.
⇨ Select suitable microscope filters.

3.6 IMAGE ACQUISITION

3.6.1 Aim

Different parameters are defined to describe and quantify the DNA damage.

Each slide should be duplicated for each condition, and 50 to 100 cells should be scored on each slide.

⇨ The comet analysis system requires a good-quality, sensitive CCD and image analysis software. Specific software programs for the comet application have been developed.

3.6.2 Parameters

3.6.2.1 Tail length

This parameter measures the migration distance of the DNA fragments from the nucleus. It increases linearly with the dose, but in highly damaged DNA, the proportion of DNA in the tail is increased but not the length.

⇨ This parameter is suitable over a restricted range of doses.

3.6.2.2 Tail moment

This parameter is defined as the product of the percentage of DNA in the tail and the distance measured from the center of the head and the mean position of the tail (Figure 3.5).

⇨ The comet software analyses the fluorescent profile of the cells. It determines the position of the comet head and tail, their respective center of mass, and their respective fluorescence intensity.

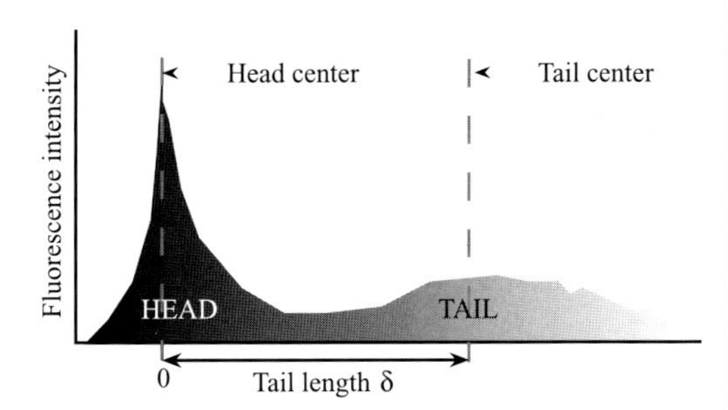

Figure 3.5 Tail moment.

3.6.2.3 Inertia moment

This parameter was introduced to measure precisely the distribution of DNA fragments in the comet tail. Cells with identical tail moment may have different inertia moment, depending on the migration of the fragment in the tail.

⇒ Usually there is a good correlation between the two parameters.

⇒ The inertia moment seems to be more sensitive.

3.6.2.4 Comet moment

This parameter does not differentiate the head and the tail of the comet. This may be useful when the DNA is highly damaged and the demarcation between the head and tail is not obvious. This method conforms more rigorously to the mathematical definition of the moment of inertia of a plane figure.

⇒ The two methods (comet moment and tail moment) produce comparable values ($r^2 = 0.89$). However, concerning low doses, the comet moment would be the parameter of greater sensitivity.

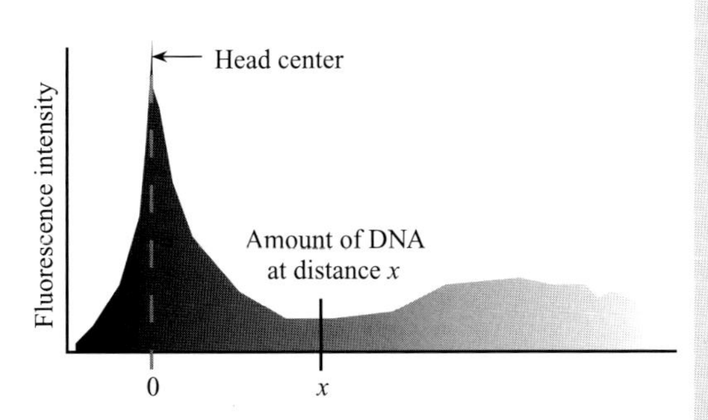

$$Comet\ Moment = \frac{\sum_0^n (amount\ of\ DNA\ at\ x)*(x)}{total\ DNA}$$

Figure 3.6 Comet moment.

3.6.2.5 Other parameters

- Diameter of the comet head.
- Head fluorescence intensity.

⇒ Those parameters may be useful to quantify the nucleus DNA. They could identify the cells in the S phase during cell cycle analysis.

- Head over tail intensity ratio.

⇨ The fluorescence intensity in the tail is correlated with the number of breaks. Nevertheless, the chromatin structure may vary with the cell cycle. Consequently, the migration features of the DNA may also vary. Alkylating agents may also impede the DNA migration in the gel. In these latter cases, the head fluorescence parameter may be useful.

3.7 APPLICATIONS

3.7.1 Study of DNA Damage in Cellular Suspension

⇨ See upper protocol.

- The cells are first incubated with the toxic agent in the culture medium and then are included in the agarose.

⇨ The incubation time of the product of interest must be optimized. Repair may occur, limiting the visible damage.

- The cells to be irradiated (ionizing or ultraviolet radiation) must be diluted in PBS or culture medium without antioxidant or photosensitizers.

⇨ The cells should be kept at +4°C, during and after irradiation, to limit the activity of the repair enzymes.

- Studies are usually performed on nonsynchronized cells. This parameter should be taken into consideration for repair experiments.

⇨ The head fluorescence can discriminate between the cells in G1 phase or in S or G2/M phase.

3.7.2 Study of DNA Damage in Monolayer Cultures

- Trypsination, or scrapping, may alter the cells. If possible, the stress should be applied after the inclusion of the cells in the agarose layer.

⇨ In a variation of the assay, the cells may be cultured in agarose for a few hours before application of the toxic agent. Single-strand break repair is very fast (half-life is about 13 min). Treatment of the cells in the gel gives more precise results for the analysis of strand breaks and for repair studies.

- Different types of radiation (ionizing, ultraviolet A or B) and toxic agents have been used.

⇨ There is a potential interest in the antenatal diagnosis of defects in repair enzymes.

UVA 1 J-cm^2 UVA 4 J-cm^2

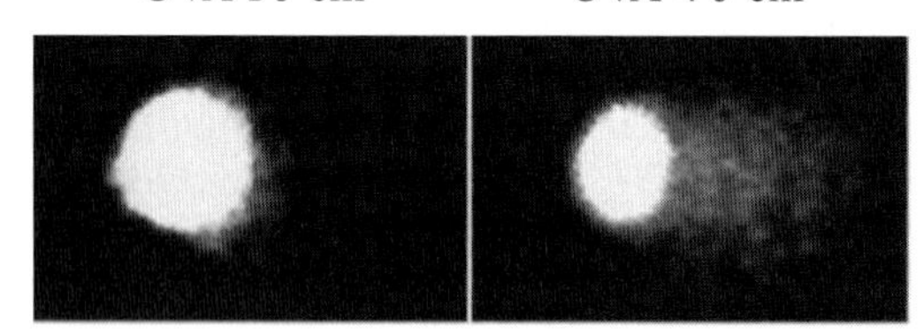

⇨ In the cosmetology field, there is a potential interest in the determination of protection coefficient of sunscreen against DNA damage.

Figure 3.7 Irradiated human fibroblast.

3.7.3 Blood

- All the nucleated circulating cells are suitable for the comet assay. Lymphocyte separation is possible but not obligatory.
  5 µL of whole blood are diluted in 110 µL of 0.6% low-melting-point agarose and spread over the first agarose layer.
- Clinical applications (prognosis and diagnosis)
- This assay may be used to study the DNA damage in patients with genetic defects in repair enzymes.

⇨ The DMSO in the lysis buffer limits the oxidation due to the heme iron.

⇨ Potential interest in the study of radio- and chemoresistance in cancerology

⇨ Potential interest in the study of secondary pathology related to oxidative stress

3.7.4 Tissues

This assay may be performed on slides prepared with fresh tissues (liver, lung, spleen, kidney, and bone marrow).

⇨ This application is limited by the tissue preparation. This must not create DNA damage.

3.8 ASSAY VARIANTS

3.8.1 Enzymatic Digestion

3.8.1.1 Aim

The DNA enzymatic digestion by repair enzymes allows targeting specific DNA lesions. Digestion occurs *in situ* on isolated DNA. The modified base excision leads to the creation of new breaks. The enhancement of the tail moment value due to the enzyme digestion is correlated to the amount of specific DNA damage.

⇨ Mainly, two enzymes that cut the *N*-glycosidic bond between the base and the 2-deoxyribose and that possess lyase activity are used: endonuclease III and the formamidopyrimidine glycosylase (FPG). They recognize oxidized pyrimidines and oxidized purines, respectively.

3.8.1.2 Protocols

- After the lysis step, the slides are equilibrated in the enzyme buffer.
- Place 100 to 150 µL of the enzymatic dilution and cover with a coverslip.

⇨ The appropriated enzyme dilution must be determined first by testing different digestion conditions.

- Incubate in the dark for a specific time (e.g., 45 min) at a specific temperature (25 or 37°C).

Control slides are incubated with the buffer without the enzyme.

⇨ The experimental conditions should be standardized to ensure reproducibility.

3.8.2 Study of DNA Repair

3.8.2.1 Aim

Follow-up of the tail moment at different times after the induction of genotoxic stress gives an idea of the cellular repair ability for the DNA damage.

3.8.2.2 Protocols

After the stress induction, the cells are maintained in standard culture conditions. The cells are then embedded in agarose and lysed at different times after the stress induction.

⇨ For adherent cells, a repair kinetic may be performed. In this assay, the cells may be included in the agarose equilibrated with the culture medium. During the repair time, cover the slides with the medium to prevent them from drying.

3.8.3 Apoptosis

3.8.3.1 Aim

During the latest steps of apoptosis the DNA is fragmented into nucleosomes by endonucleases.

3.8.3.2 Protocols

The nucleosomes (multiple of 200 base pairs) migrate through the gel during the alkaline electrophoresis step, whereas the nuclear DNA disappears. The proportions of these visually highly damaged cells are manually counted.

⇨ This approach does not replace the DNA ladder technique. Nevertheless, it is very useful to score apoptotic cells in a cellular population.

3.9 CONCLUSIONS

The comet assay has many potential applications in biology and in the medical field. Only few cells are required. Nevertheless, the technique is not standardized among the different laboratories, and this impedes the comparison of the results obtained. The use of repair enzyme digestion enhances the specificity of the technique and gives access to families of modified bases. The analytical techniques based on the separation of modified bases are yet necessary to perfectly identify the DNA lesions created.

3.10 ACKNOWLEDGMENTS

We thank Jean Cadet and all the members of his laboratory (Département de Recherche Fondamentale sur la Matière Condensée, Scib/Lan, CEA Grenoble, France). We also thank Alain Favier (Laboratoire Biologie du Stress Oxydant, University Joseph Fourier) for very valuable assistance.

Chapter 4

High-Resolution FISH Mapping

Karine Monier and
Sophie Rousseaux

CONTENTS

4.1 HIGH-RESOLUTION PHYSICAL MAPPING USING FLUORESCENT *IN SITU* HYBRIDIZATION (FISH) ON DECONDENSED DNA FIBERS

4.1.1 Purpose

Physical mapping of genomes is a basic approach for the analysis of genetic abnormalities.

⇨ It enables the identification of deleted or altered regions of the genome in diseases associated with alteration of the genome. It therefore contributes to the identification of genes involved in the corresponding phenotypes.

Fluorescent *in situ* hybridization (FISH) is a tool of choice for *in situ* mapping.

⇨ FISH allows the direct visualization of DNA sequences on metaphase chromosomes or interphase nuclei using fluorescent probes.

The resolution of FISH mapping depends on the extent of DNA relaxation in the analyzed samples.

⇨ The resolution is the minimal distance in kilobase (kb) between two sequences along the DNA fiber, which allows them to be ordered.

⇨ This distance is 1 to 2 megabases (Mb) on metaphase chromosomes, and 50 to 100 kb in interphase nuclei.

New methods allowing *in situ* hybridization on artificially decondensed DNA will fill the gap between the molecular biology and the *in situ* approaches.

⇨ Indeed FISH mapping on decondensed DNA allows the discrimination of distances of a few kb.

4.1.2 Methods to Obtain Decondensed DNA Fibers

Preparations from whole cells allow the maintenancc of a residual nuclear structure. Preparations from DNA in solution or high-molecular-weight DNA, extracted from cells lysed in agarose blocks, are used to obtain oriented or linear DNA fibers (Figures 4.2 and 4.3).

⇨ Halo methods (Figure 4.1)

⇨ Oriented fibers (Figure 4.2)

⇨ These procedures use chemical treatments (high ionic strength, detergents, soda, etc.) which result in a controlled or total cell lysis, dissociate the nucleoprotein complexes, and free the DNA fiber.

In some cases (oriented and linear fibers), a mechanical treatment is applied after the chemical procedures, in order to stretch the DNA fibers.

The unidirectional orientation of the fibers, as well as the homogeneity of the level of decondensation, are highly variable depending on the method.

⇨ According to the method, the decondensation level varies between 0.014 and 0.5 µm/kb.

0-8493-0817-8/01/$0.00+$1.50

4.1.3 Preparation of Relaxed DNA Fibers by the "Halo" and Related Methods

Two methods allow the preparation of relaxed DNA fibers from living cells, while preserving a residual nuclear structure.

These methods are easy to apply. The cells are either grown directly on the slides or applied by centrifugation. The glass slides are then immersed in different solutions.

The "halo" method (technique 1) relies on the treatment of the cells with high salt concentrations in the presence of nonionic detergents.

⇨ The "halo" method
⇨ A method using a topoisomerase II inhibitor and an alkaline treatment.
⇨ The decondensation levels of the DNA fibers obtained by these two methods are different.

⇨ The word "halo" is used to describe the nuclear structures, which are characterized by 100 to 200 kb DNA loops attached to a residual nuclear structure.
⇨ The halo method allows a condensation level close to the one of the DNA double helix in its B conformation (0.34 µm/kb).

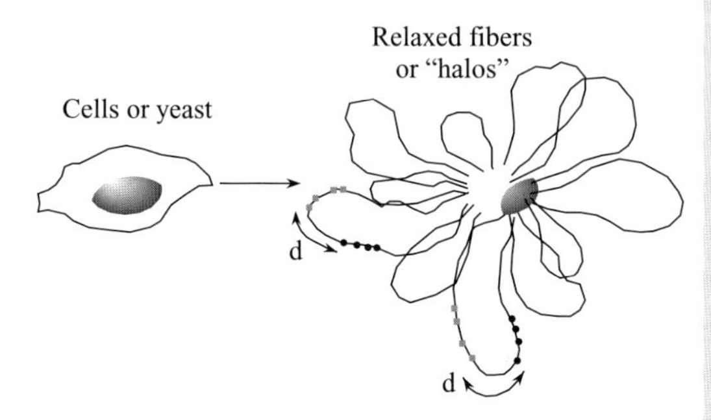

Figure 4.1 Relaxed DNA fibers or "halos." The treatment of cells or yeast at high salt concentration with detergent frees DNA loops. These more or less relaxed DNA loops stay anchored to the residual nuclear structure.

The second method (technique 2) relies on the combination of a topoisomerase II inhibitor and an alkaline treatment. The free DNA loops remain attached to a residual nuclear structure.

These two procedures maintain some spatial cohesion between the loops of the same nucleus. In most cases, the signals can be analyzed in each individual nucleus, and the number of alleles and their structure are possible to determine.

The attachment of DNA loops onto residual nuclear structures was also used as an indirect approach to study chromatin organization within the interphase nucleus.

⇨ This method does not allow a decondensation level as high as the halo method (0.014 µm/kb).

⇨ However, the attachment of some DNA regions on the residual nuclear structure constitutes an obstacle for the mapping of these regions and makes distances difficult to measure on bent fibers.

4.1.4 Preparation of Oriented DNA Fibers

A total lysis of the cells is required in order to prepare oriented DNA fibers. This can be performed directly on the slide from a suspension of fixed cells (technique 3) or unfixed cells (techniques 4 and 5). Alternatively, cells embedded in agarose blocks can be used (technique 6).

⇨ The methods of cell lysis on the slides are based on a treatment with soda (technique 3) or the combined use of SDS, Triton, and EDTA (technique 4) or salt, SDS, and Triton (technique 5). The orientation of fibers is obtained by draining the solutions on tilted slides.

➾ The cell lysis performed in agarose uses *N*-lauroylsarcosine and EDTA in the presence of proteinase K (technique 6), and the molecules are mechanically stretched with the edge of a second slide.

These methods produce longer DNA fibers than the methods described previously.

➾ Indeed, the DNA fibers, freed from the nuclear body, may reach a length up to 1 Mb (technique 5).

In all these preparations, the orientation of the fibers makes the distance measurements easier to perform. It also allows the mapping of the entire genome, due to the absence of residual nuclei.

➾ However, the analysis of individual cells/nuclei is impossible.

The mean decondensation levels obtained with these approaches are similar to those obtained with the halo method.

➾ They are highly variable from one fiber to another.

➾ The frequency of breakage increases with the size of the molecules, which makes it difficult to analyze sequences of several hundreds of kb apart.

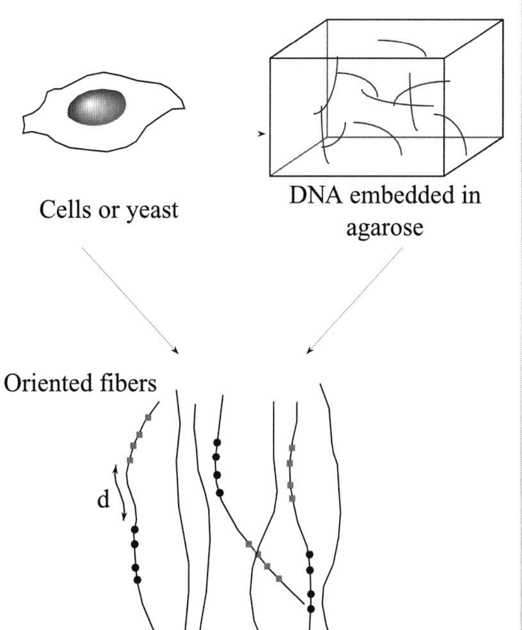

Figure 4.2 Oriented fibers. Cells or yeast are lysed, either on the slide or in an agarose block and then mechanically stretched to obtain fibers freed from their nuclear anchorage. These fibers are oriented in one predominant direction. The level of DNA decondensation varies greatly from one fiber to another.

4.1.5 Preparation of Linear or Combed DNA Fibers

The preparations of linear DNA fibers are performed using high-molecular-weight DNA extracted from agarose embedded cells, or with DNA purified in solution. This technique uses specifically pretreated glass slides onto which the DNA molecules spontaneously anchor by their ends.

➾ The molecule anchoring is performed by incubating the pretreated glass slides in a DNA solution. The stretching of these molecules is obtained by applying a pulling force generated by the movement of the meniscus at the water–air interface when the slide is pulled out of the DNA solution at a constant speed. This method was named molecular combing.

The main advantage of molecular combing is that it generates DNA fibers with a more constant and reproducible level of decondensation than other methods.

⇨ The combed molecules of different origins (total genomic DNA, yeast artificial chromosomes, cosmids, lambda phages) are linear and decondensed in a homogeneous and reproducible manner.
⇨ Molecules of a maximal length of 500 to 700 kb can be stretched without the risk of breakage being too high.
⇨ This decondensation is approximately 0.5 μm/kb, which corresponds to a higher stretching than the DNA in its B conformation (increased ×1.5).
⇨ *See* Figures 4.3 and 4.4.

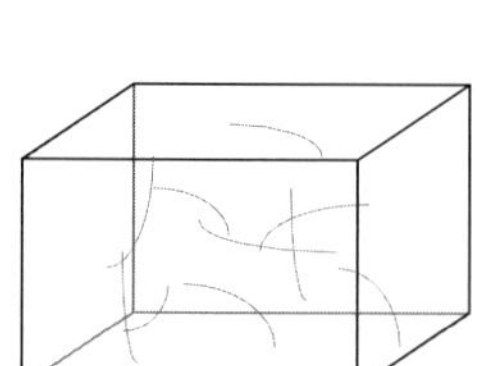

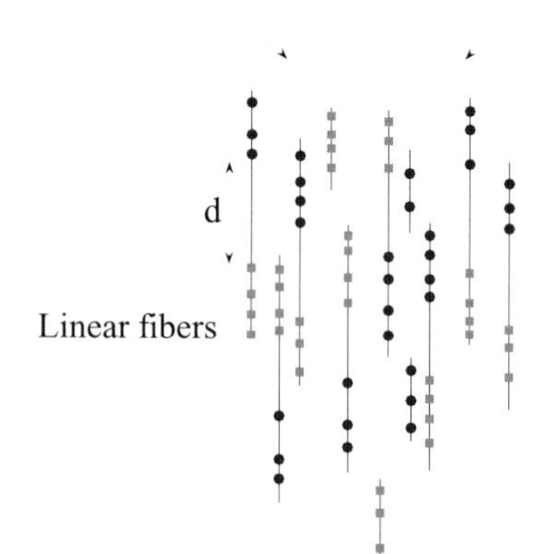

Figure 4.3 Linear DNA fibers or "combed" fibers. Purified DNA molecules, embedded in an agarose block or in solution, are mechanically stretched after anchorage by their ends to an appropriate surface. This results in DNA fibers that are perfectly aligned in the same direction. The decondensation level is homogeneous.

4.1.6 Applications

The methods described above allow the study of nucleic acid sequences distribution with a high resolution.

⇨ A high degree of decondensation is recommended to order sequences which are very close to each other, or which partially overlap.
⇨ A low degree of decondensation is more suitable for the ordering of large sequences (YAC clones) or of those several hundreds of kb apart.

The detection threshold of the method depends on the intensity of the signal and on the sensitivity of the observation system.

The intensity of the signal is related to the nature of the fluorochrome and to the signal amplification method.

⇒ The detection threshold of the technique is the minimal size in kb required for a sequence to be visualized as a dotted signal

⇒ This amplification can be obtained by enzymatic systems or by successive antibody layers.

⇒ With the molecular combing method, 1-kb fragments can be visually detected using three successive layers of antibodies.

High-resolution mapping techniques on decondensed DNA have tremendously increased the resolution of FISH mapping on genomes.

⇒ Most of these methods allow the detection of sequences of a few kb and the discrimination of sequences less than 5 kb apart.

⇒ They have been applied to order DNA fragments cloned in YAC and cosmids as well as restriction fragments.

These techniques have also allowed the study of genetic alterations, amplifications, and deletions of genes.

The direct conversion of a distance measured in μm into a length in kb is only possible if the decondensation level is reproducible.

⇒ This requirement gives to the method of molecular combing a tremendous advantage compared to the other methods (Figure 4.5).

4.2 FISH PROTOCOL ON MOLECULAR COMBED DNA

Molecular combing of target DNA molecules combined with FISH represents a major methodological improvement for high-resolution physical mapping of the genome.

⇒ FISH on molecular combed DNA allows the precise and reproducible measurement of hybridization signals from DNA probes.

Standard FISH procedures can be successfully applied on combed DNA.

⇒ Optimized hybridization conditions allow higher hybridization efficiencies as well as more reliable and accurate measurements, even when using small-size probes.

4.2.1 Optimization of FISH on Combed DNA

4.2.1.1 Target DNA

DNA molecules of several origins and sizes can be used as targets.

⇒ Phage and cosmid DNA are extracted in solution.

⇒ Large size DNA (>400 to 500 kb), such as YAC and total genomic DNA, should be prepared from cells embedded in agarose blocks as for pulse field gel electrophoresis.

The optimal size of regions to be mapped should be within 1 to 200 kb.

⇒ Fragments up to 1 Mb have been successfully combed, but the proportion of broken molecules is high.

Density of DNA molecules on the surface should be adjusted by varying DNA concentration in the solution.

⇨ Usually it reaches 50 to 100 molecules of lambda DNA or 20 to 50 molecules of YAC DNA per microscopic field of view of 100 × 100 µm (objective lens 100×).

4.2.1.2 DNA probes

Many types of probes with different sizes can be hybridized on combed DNA.

⇨ For example, these probes can be plasmid, lambda phages, cosmids, BAC, P1, YAC DNA, or cDNA probes or total genomic DNA.
⇨ Probes as small as 1 kb or as long as several hundreds of kilobases have been detected by FISH on combed DNA.

Probes can be labeled either by nick translation or random priming.

⇨ The same hybridization efficiency is obtained with both labeling methods.
⇨ Purification of the labeled probes from nonincorporated nucleotides does not improve the signal/noise ratio.
⇨ For some yet unknown reasons, signals generally appear as "beads on a string," and this pattern is observed whatever the size of the probe.

4.2.1.3 Hybridization conditions

In order to reduce the background noise, it is recommended to quench nonspecific anchorage sites by preincubating the slides in a blocking solution before hybridization.

Accurate and reproducible results are obtained when denaturation of the slides is performed by dipping the slides into the denaturation solution.

⇨ Indeed, when the slides are denatured in an oven, the result is a less homogeneous and a less reproducible hybridization.

When a genomic probe is used, competitor DNA is required at an appropriate concentration in order to suppress nonspecific hybridization signals due to repeated sequences.

⇨ However, it should be kept in mind that the quenching of large blocks of repeated sequences might create interruptions of the hybridization signals along the molecules, and care should be taken when analyzing fragments enriched in repeated sequences

The post-hybridization wash conditions of temperature and stringency should be chosen to minimize background noise while maintaining a high hybridization rate.

⇨ With a lambda probe hybridized on itself, post-hybridization washes (3 × 5 min in both 50% formamide/2× SSC, 2× SSC) performed at 37°C allowed 30 to 50% hybridization efficiency, whereas the same washes performed at room temperature (RT) increased the efficiency rate to 100%. However, the background noise increases when washes are performed at RT.

4.2.1.4 Detection of signals

The hybridization signals cannot be detected by eye without amplification.

Several double-color signal amplification protocols have been successfully applied, as described in the following protocol section (see below).

Hybridization signals appear with a higher intensity when five layers of proteins are applied, but the background noise is also higher than with two or three layers of proteins.

⇨ A 5-layers detection is recommended when using large-size probes (YACs and cosmids).

⇨ A 3-layers system is more suitable with smaller probes (cDNAs or plasmids).

Enzymatic amplification using the tyramide system is very efficient in increasing the signal intensity without generating background noise.

⇨ The tyramide system does not allow a polychromic detection but can be very useful to detect small-size probes.

Unfortunately, we did not succeed in counterstaining combed DNA molecules after hybridization with the usual fluorescent dyes. However, the whole combed DNA can be used as a probe to counterstain the fibers (Figure 4.4) providing that the number of probe molecules to be mapped is in excess (at least 50 times) of the probe used as a counterstain.

4.2.1.5 Image analysis and measurements

When using several probes in double-color FISH, the accuracy of the measurements relies on a precise correction of the shift generated by switching the filters specific for each wavelength.

⇨ An additional double-labeled probe can be hybridized in the region of interest as a standard to check this shift.

⇨ Use of a double-band pass filter or an automatic filter wheel are other ways to overcome the problem.

The accuracy of measurements on digital images may be improved by applying an image restoration procedure based on iterative constrained deconvolution.

⇨ The procedure of deconvolution removes the blur of fluorescence.

⇨ Using this method, the signals of two contiguous cDNA probes of respectively 1.2 and 1.9 kb in size were resolved in 75% of analyzed molecules.

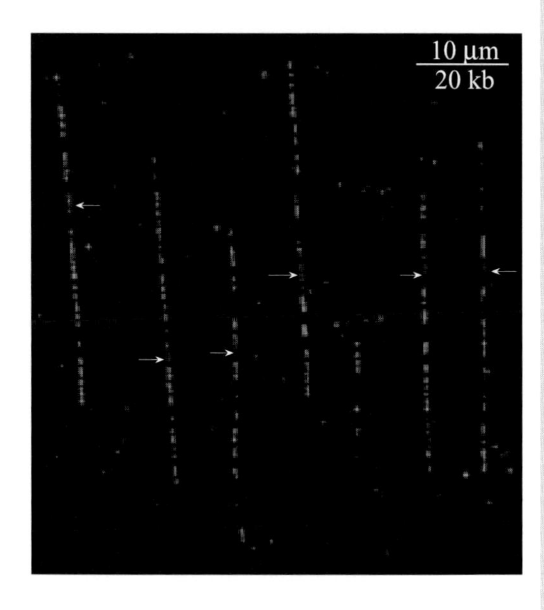

Figure 4.4 Examples of double-color hybridization of a plasmid probe on a region of interest cloned in a lambda phage, stretched with the method of molecular combing. The plasmid probe of 2.5 kb is visualized in red (arrows), and the phage lambda DNA molecules of 50 kb are visualized in green. The molecules are all oriented in the same direction, and the sizes of the signals are homogeneous. The position of the 2.5 kb fragment is reproducible from one molecule to another (variation: 10%). It is 18 kb from the closest end of the phage molecule and 29.8 kb from the farthest. Probes were labeled with biotin or digoxigenin, and the signals were detected with two or three layers of proteins coupled with fluorochromes.

4.2.2 Protocol of FISH on Combed DNA

4.2.2.1 Slides of combed DNA (target)

The target DNA is combed on 22 × 22 mm silanized coverslips.

⇨ These surfaces are 22 × 22 mm coverslips mounted onto ordinary slides with glue.

Dehydrate the slides.

⇨ Dehydration is carried out through an ethanol series (70, 90, and 100%, 5 min each).

Bake them in an oven at 60°C for 3 h.

⇨ If not used immediately, the slides may then be stored in a desiccated box at –20°C.

Before the FISH experiment, incubate the slides at 37°C for 30 min with a blocking buffer containing 3% BSA/2× SSC, rinse them in 2× SSC, and dehydrate them again.

For denaturation, immerse the slides in a 70% formamide/2× SSC pH 7 solution at 70°C for 2 min and immediately dehydrate them in a cold ethanol series.

4.2.2.2 Probes

Precipitate the labeled probes with 10 µg of salmon sperm and if necessary cot-1 DNA (in the case of competitive hybridization).

⇨ Amount of probes to be used for each slide: 100 ng of cosmid-, lambda-, or c-DNA probe, 300 ng of YAC probe.

Resuspend them in 10 µL of hybridization buffer.

⇨ The hybridization buffer is made of 50% formamide/2× SSC/10% dextran sulfate/0.1% Tween 20.

Denature probes at 75°C for 5 min.

In the case of suppressive hybridization, preincubate at 37°C for 30 to 60 min.

4.2.2.3 Hybridization

Apply 10 µL of the probe mixture on each slide.

Seal with a coverslip.

Incubate at 37°C overnight in a moist chamber.

4.2.2.4 Post-hybridization washes

Wash slides with agitation in 50% formamide/2× SSC, pH 7 at RT (3× 5 min).

Repeat the same washes in 2× SSC at RT (3× 5 min).

4.2.2.5 Detection

Incubate them at 37°C for 30 min with a blocking buffer containing 3% BSA/4× SSC/0.1% Tween 20 and rinse in 2× SSC.

Incubate the slides at 37°C for 30 min with a detection buffer (1% BSA/4× SSC/0.1% Tween 20) containing the detection protein coupled with the fluorochrome at the appropriate dilution.

Wash the slides with agitation in 4× SSC at RT (3× 5 min).

Repeat the last two steps for each of the successive layers of detection proteins.

At the end of the detection procedure, mount the slides with an antifade buffer (90% glycerol, 2.3% DABCO, 20 m*M* Tris-HCl pH 8).

⇨ As an example, two probes labeled respectively with biotin-dUTP and with digoxigenin-dUTP can be co-detected in double color with the following five-layers amplification system (Figure 4.5):

1. Avidin-TRITC 1/100 (Sigma) and mouse anti-digoxin antibody labeled with FITC 1/50 (Jackson)
2. Biotinylated anti-avidin antibody 1/100 (Vector) and donkey anti-mouse IgG antibody labeled with FITC 1/50 (Jackson)
3. Avidin-TRITC 1/100 and mouse anti-rabbit IgG antibody labeled with FITC 1/50 (Jackson)
4. Biotinylated anti-avidin antibody 1/100 and rabbit anti-FITC antibody 1/200 (Cambio)
5. Avidin-TRITC 1/100 and goat anti-rabbit IgG antibody labeled with FITC 1/50 (Cambio)

Figure 4.5 Examples of fluorescent *in situ* hybridization of cosmid probes (35 to 40 kb inserts) on YAC DNA molecules containing the region to map and stretched with the molecular combing method. The technique was applied here to order some probes. The green signal corresponds to the same genomic fragment in both boxes (1), whereas fragment 2 (in right box) is closer to fragment 1, than fragment 3 (in left box). The order of the fragments (probes) is therefore 1, 2, and 3. The homogeneity of the level of decondensation of combed fibers enables the precise measurement of distances in kb between the two probes (variations <5%): in the upper box, the hybridization signals are 80 μm apart (which corresponds to a 160 kb distance), whereas in the lower box the distance between the two signals is 95 μm (which corresponds to 190 kb). Probes were labeled with biotin or digoxigenin, and the signals were detected with five layers of proteins coupled with fluorochromes (*see* Protocol).

Chapter 5

Fluorescence Imaging

Yves Usson

CONTENTS

5.1 PRINCIPLE OF FLUORESCENCE IMAGING

❑ *Advantages*

The main advantage of fluorescence imaging is to offer the possibility of associating a given fluorophore with a particular molecule. Thus, it is possible to label specifically and simultaneously different targets with different fluorochromes that can be easily discriminated on the basis of their emission wavelengths.

Fluorescence imaging is particularly well suited for the study of the intracytoplasmic or intranuclear distribution of molecules. It is also the method of choice for investigating *in situ* molecular colocalization.

⇨ Simultaneous specific labeling of different targets

⇨ With a conventional fluorescence microscope, it is possible to image and discriminate up to three different fluorochromes.

❑ *Disadvantages*

Fluorescence imaging also has some weaknesses when compared to conventional bright-field microscopy.

⇨ Poor spatial resolution: the images are corrupted by out-of-focus haze and glare.

The special optical requirements of fluorescence are responsible for a substantial increase in the cost of the microscope parts. For instance, it is mandatory to have multiple filter sets to cover the range of the different fluorophores currently available.

⇨ Expensive optical parts: interferential filters, expensive light sources with a short lifetime (200 to 300 hours), specially processed lenses for a broader light spectrum.

Quantification of molecules based on the measurement of fluorescence intensity is made very difficult by the multiplicity of error sources due to fluorophore properties and to the incoherent nature of fluorescence emission.

⇨ Measuring significant fluorescence intensities is tricky; one has to face image blur, fading, spectral shift, leakage, quenching, etc.

5.2 FLUORESCENCE MICROSCOPES

5.2.1 Basic Principle

Fluorescence microscopy relies on the properties of the fluorochromes. These particular molecules absorb the energy carried by photons in a given band of wavelengths and give back a fraction of this energy by emitting new photons in a band of longer wavelengths.

⇨ *See* Figure 1.1.

0-8493-0817-8/01/$0.00+$1.50

In the mid-1950s, the first fluorescence microscopes were based on a very simple design directly derived from that of classical transmission microscopes. The conceptual differences with the latter concerned the type of light source, the use of a very selective filter (excitation filter) in the condenser system, and finally the addition of a second filter after the objective lens (emission filter).

➩ The light source is an arc lamp that provides a broad spectrum with high intensities.

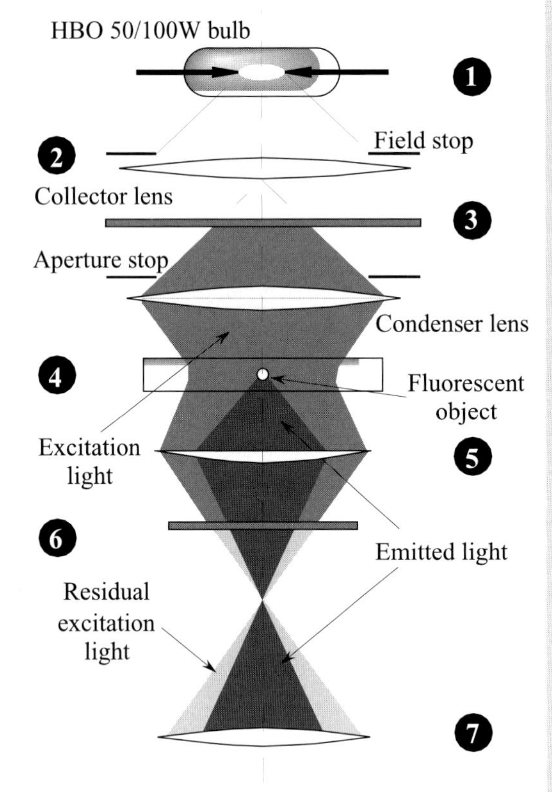

1. **Light source: mercury or xenon arc lamp that delivers a high intensity over a wide spectrum**
2. **Condenser: focuses the light source on the specimen**
3. **Excitation filter: a short-pass or band-pass filter with cutoff wavelengths chosen to select the adequate excitation wavelength**
4. **Specimen**
5. **Objective lens: collects the light from the specimen**
6. **Emission filter (stop filter): a long-pass or band-pass filter with cutoff wavelengths chosen to reject the excitation light and let the emitted light pass through**
7. **Eyepiece**

Figure 5.1 Schematic diagram of a linear fluorescence microscope.

However, this linear organization of optical parts had many drawbacks.

As a matter of fact, the images obtained by such a system were hardly satisfactory. The main problem came from the huge difference in intensity between the amount of light required to properly excite the fluorophores and the amount of light emitted by them. Even though a selective emission filter was used to reject the excitation light, a significant amount of it could still reach the image plane and impair the contrast of the final image.

➩ The intensity of emitted light can be 10,000 times lower than that of excitation light.

➩ If the emission filter attenuates the excitation light by a factor of 1000 times, it remains higher than the emitted light.

5.2.2 Epi-Illumination Microscope

5.2.2.1 Principle

In the early 1960s, in the Netherlands, J. B. Ploem invented a new type of fluorescence microscope. The design was based on the concept of epi-illumination. The idea behind it is to use the objective lens for two different purposes:

- As an imaging lens (its usual role)
- As an illumination lens; in this case, the objective lens behaves like the condenser lens that is used to focus the light source on the specimen.

In fact (*see* Figure 5.2), the light source is deported on the side of the microscope and the excitation light rays are sent on a dichroic mirror, oriented with an angle of 45°, which redirects them toward the back entrance pupil of the objective lens. The latter focuses the excitation rays onto the specimen. After traversing the object plane the excitation rays diverge and exit from the microscope.

⇨ A dichroic mirror is a mirror with a special coating that makes it possible to reflect or transmit light as a function of the wavelength. When the light rays come with an incidence angle of 45°, they are either reflected, if their wavelength is shorter than a given cutoff wavelength, or transmitted if their wavelength is longer.

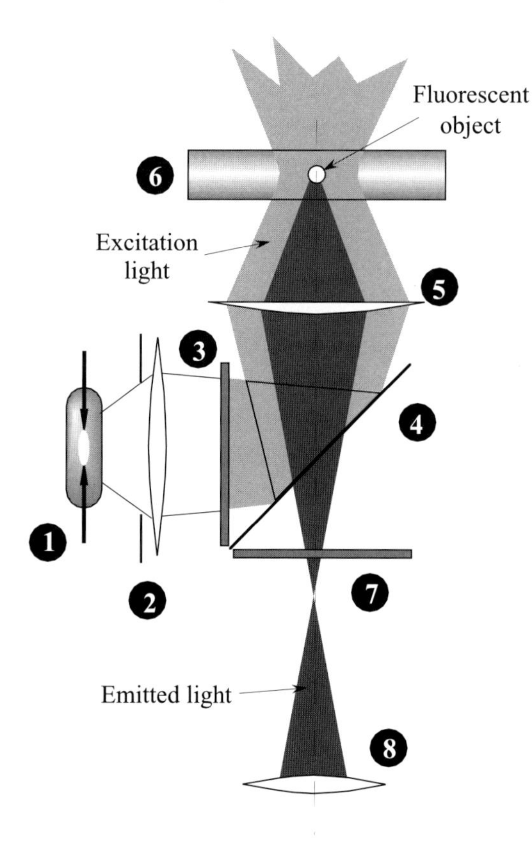

1. **Light source: mercury or xenon arc lamp that delivers a high intensity over a wide spectrum**
2. **Condenser**
3. **Excitation filter: a short-pass or band-pass filter with cutoff wavelengths chosen to select the adequate excitation wavelength**
4. **Dichroic mirror: behaves like a beam splitter; it reflects the excitation light and lets the emitted light pass through**
5. **Objective lens**
6. **Specimen**
7. **Emission filter: a long-pass or band-pass filter with cutoff wavelengths chosen to reject the excitation light and let the emitted light pass through**
8. **Eyepiece**

Figure 5.2 Schematic diagram of an epi-fluorescence microscope.

A part of the emitted fluorescent light is collected by the objective lens and sent to the dichroic mirror. Because the wavelength of the emitted light is longer than that of the excitation light, it is transmitted through the mirrors and eventually reaches the eyepiece.

5.2.2.2 Image formation

In an ideal system, the image of a single point source is expected to be a single point too. However, during the imaging process by a real objective lens, small amounts of light are diffracted and do not reach the image plane at the position where they should converge. These diffracted rays interfere, either in a constructive or in a destructive way, and give rise to interference rings (*see* Figure 5.3).

The mathematical representation of this phenomenon is called the point spread function (PSF). This function is used to describe the relationship between a real object and its image formed in the microscope. The general expression of the image formation is:

➾ In other words, the PSF describes how the objective lens of the microscope smears the light coming from the point source around the position of its ideal point image.

$$i_{xy} = h_{uv} \otimes o_{xy}$$

➾ $\otimes$ denotes a two-dimensional convolution operator. That is:

$$i(x, y) = \int_{-\infty}^{\infty}\int_{-\infty}^{\infty} [h(u, v)o(x-u, y-v)]dudv$$

where i_{xy} is the light intensity distribution in the image plane, o_{xy} the light intensity distribution in the focal plane, and h_{uv} is the PSF.

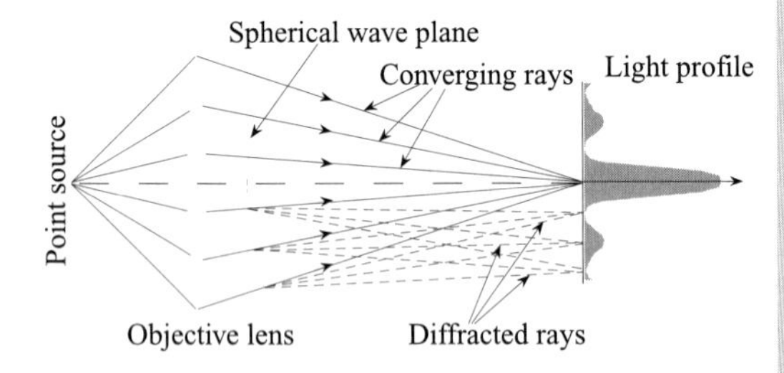

Figure 5.3 Image formation by an objective lens.

Actually, in a conventional microscope two different PSFs are involved in the imaging process:

1. The 2-D in focus point spread function h_i
2. The out-of-focus point spread function h_o (defocus blurring function)

The in-focus PSF is the mathematical description of the Airy disk image formed by the objective of an ideal, dimensionless point object on the optical axis of the focal plane.

This familiar image (Figure 5.4), of a central maximum surrounded by rings of weaker intensity, results from the inability of the lens to collect light scattered through angles greater than α, the maximum acceptance half-angle of the objective (highest spatial frequencies).

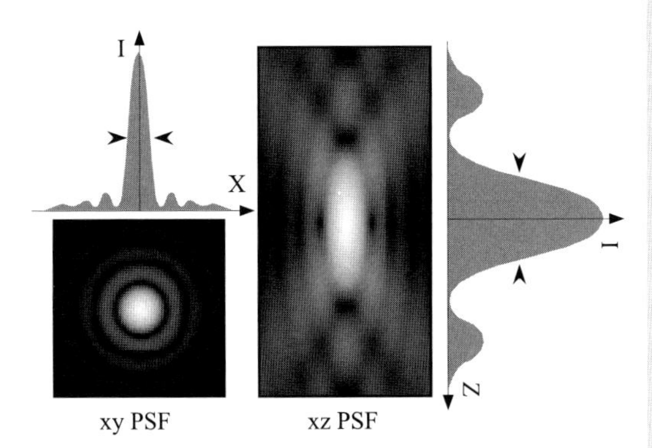

Left, lateral PSF (*xy*) and intensity profile
Right, axial PSF (*xz*) and intensity profile

The arrows on intensity profiles show the lateral and axial FWHM (full width half maximum), which correspond to the lateral and axial resolving power, respectively.

Figure 5.4 Theoretical point spread functions in a conventional epi-fluorescence microscope.

The numerical aperture NA is defined as:

$$\mathrm{NA} = n \sin \alpha$$

α being maximally 67.5° for an oil-immersion objective with an NA of 1.40, used with immersion oil of refractive index n = 1.515.

The resolution of an image obtained in an optical microscope is diffraction limited. Rayleigh's criterion for the resolution limit of the optical microscope says that d_{min}, the distance between two point objects at which the peak intensity of the Airy disk of the first object overlaps with the first minimum of the Airy disk of the second, is defined by:

$$d_{xy} = \frac{0.61\lambda}{\mathrm{NA}}$$

λ is the wavelength of light.

This definition assumes that the objective is receiving light from a condenser of at least equal aperture. In epi-fluorescence and reflection microscopy, of course, the objective also serves as a condenser, satisfying this condition.

The in-focus PSF, h_i, thus describes the inevitable blurring present in the image i, at the primary image plane of the objective, of an infinitely thin in-focus object o. Using aberration-free conventional optics it leads to the general formula of image formation:

$$i_{xy} = h_{i_{uv}} \otimes o_{xy}$$

Similarly, the out-of-focus PSF, h_o, defines the manner in which this Airy disk image is further degraded when the idealized point object is progressively moved along the microscope optical axis away from the plane of focus.

$$i_{xy} = h_{o_{uv\Delta z}} \otimes h_{i_{uv}} \otimes o_{xy}$$

⇨ Δz is the distance of the specimen from the focal plane.

These two PSFs may be convolved to give the microscope's total three-dimensional PSF, h_{uvw}, and the imaging of a three-dimensional object may be described by the convolution of the three-dimensional object with the three-dimensional PSF:

$$i_{xyz} = h_{uvw} \otimes o_{xyz}$$

According to diffraction optics theory, for an ideal lens this three-dimensional PSF should be symmetrical above and below the focal plane.

⇨ However, it has been shown, by imaging subresolution fluorescent beads, that this is far from being the case. Not only is the PSF asymmetrical along the z axis, but it also fails to show the expected circular symmetry in planes perpendicular to z. Consequently, the three-dimensional PSF must be determined experimentally using subresolution point test objects.

It was found that the 3-D image of a subresolution sphere is elongated in the direction of the optical axis; this image has the shape of an American football. This results in severely impairing the spatial resolution of a microscope along the optical axis.

5.2.2.3 Limitations

Conventional epi-fluorescence microscopy, widely used for the localization of fluorescent probes in cells and tissues, suffers from one major disadvantage, namely, *out-of-focus* blur. The illumination of the entire field of view of the specimen with intense light at the excitation wavelength excites fluorescence emissions throughout the whole depth of the specimen. Light coming from regions of the specimen above and below the focal plane is collected by the objective lens and thus contributes to the final image as out-of-focus blur, seriously degrading it by reducing contrast and sharpness.

5.2.3 The Confocal Microscope

⇨ Out-of-focus haze, poor contrast

The answer to the drawbacks of epi-fluorescence microscopy is the confocal laser-scanning microscope (CLSM). In such a microscope, the imaging process is based on the use of a high-intensity source of light that is limited in size to a single spot.

❑ *Advantages*

Confocal scanning optical microscopy presents the following main advantages:

- Almost complete elimination of out-of-focus blur in fluorescence or reflection imaging
- Improved lateral and axial resolutions

⇨ Actually, the gain in lateral resolution is rather small (2%). The real improvement concerns the axial resolution.

- Capability for noninvasive serial optical sectioning

⇨ Cellular tomography

- Generation of high-resolution digitized data sets of the 3-D distribution of labels within cells and tissues.

⇨ Suitable for subsequent image processing

❑ *Disadvantages*

- Poor signal vs. noise ratio; only light coming from the focal plane actually reaches the photodetector. Therefore the amount of measured light is very small.

⇨ However, this small amount light is the only relevant signal to be measured.

- The acquisition time of a series of confocal section is long.

⇨ CLSM is definitively not suited for routine analysis!

- Very sensitive to photobleaching
- Generates huge data sets which require the use of powerful computers with high-capacity storage devices

⇨ Nowadays most personal computers can handle these amounts of data.

5.2.3.1 Principle of the CLSM

The fundamental idea of confocal microscopic imaging is to prevent out-of-focus photons from ever contributing to the recorded image. In the simplest case, this is achieved first by illuminating the specimen with a diffraction-limited spot of light, produced by demagnifying a laser-illuminated aperture placed in a conjugated plane of the objective lens focal plane. Those photons emitted in the volume of the illuminated focal spot of light are then detected point by point with a photodetector (PMT) during the scanning of the sample. Those photons arising from out-of-focus regions are prevented from reaching the detector by placing an imaging aperture in front of it, in a position confocal with the illuminated point within the 3-D specimen.

⇨ *See* Figure 5.5.

⇨ *See* Section 5.3.2.1.

⇨ In common with other forms of object plane scanning microscopes, an image is generated by raster scanning a given field of the specimen with the illuminating spot, digitizing the analogue output from the PMT to create the digital image pixels, and displaying the resulting digital image on a monitor.

Therefore, only light emanating from the illuminated point passes the imaging aperture (confocal pinhole) and reaches the detector. Light from regions above and below the in-focus point comes to focus elsewhere and is therefore almost totally prevented from reaching the light detector (Figure 5.5). As a consequence, fewer photons will reach the detector, and high-sensitivity light detectors must be used.

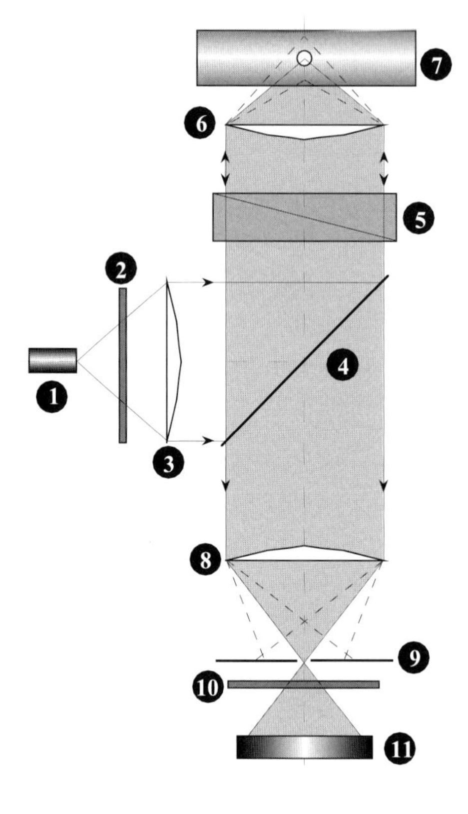

1. **Point light source (laser beam)**
2. **Excitation filter**
3. **Collector lens**
4. **Dichroic mirror (beam splitter)**
5. **Scanning device: a pair of orthogonal oscillating mirrors deflect the rays in order to scan all or part of the microscope field**
6. **Objective lens: focalizes the image of the point light source on the specimen, then collects the light emitted at its focal point**
7. **Specimen**
8. **Collimator lens: focuses the parallel light rays on the pinhole**
9. **Pinhole diaphragm: rejects "out-of-focus" rays (dotted lines) and lets the rays coming from the object focal point of the objective lens reach the PMT**
10. **Emission filter (stop filter)**
11. **Light detector (PMT): converts the collected light into an electrical signal**

Figure 5.5 Schematic diagram of a confocal microscope.

5.2.3.2 Image formation in a CLSM

There are various ways of describing the depth discrimination of a scanning optical microscope and of estimating the improvement in axial resolution in comparison with a conventional microscope. These are based on modeling and the experimental determination of the imaging of a variety of object forms and test specimens. One measure of axial resolution is to consider the decrease in the axial image intensity of an idealized point object situated on the optical (z)

axis. Using the formulae developed by Born and Wolf to define the PSF of the objective lens, in a confocal microscope with an ideally small confocal imaging aperture, the axial intensity of the image of an ideal point object can be calculated at different degrees of defocus. From this it can be shown that the depth of field of a confocal microscope, defined as the half-width of this function, is reduced relative to that of a nonconfocal microscope by a factor of about 1.4.

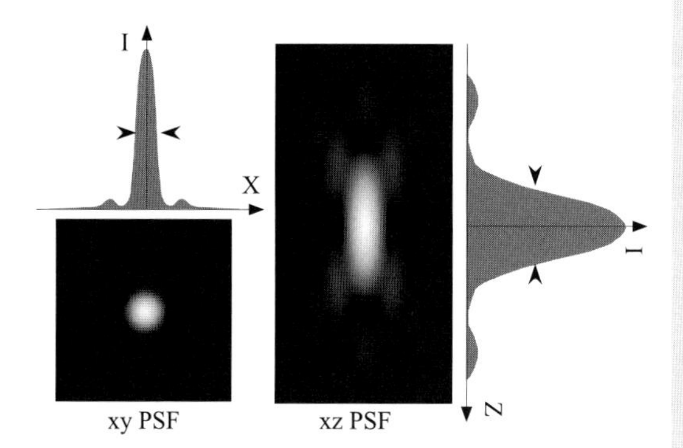

Left: lateral PSF (*xy*) and intensity profile
Right: axial PSF (*xz*) and intensity profile

The arrows on intensity profiles show the lateral and axial FWHM, which correspond to the lateral and axial resolving power, respectively.

Figure 5.6 Theoretical point spread functions in a confocal fluorescence microscope.

Theoretical considerations dictate that this confocal axial resolution will always be about three times poorer than the optimal confocal lateral resolution.

$$d_z = \frac{1.4n\lambda}{\mathrm{NA}^2}$$

Because this confocal increase in axial resolution is inversely proportional to the square of the numerical aperture of the objective, it is imperative to use objectives of the highest NA to obtain the best axial resolution.

There are other definitions of depth of field that may be of a greater practical importance in fluorescence microscopy. One of these is the variation of the integrated intensity of the image of such a point source. This indicates how well the microscope discriminates against parts of the object not in the focal plane (Figure 5.7). In conventional microscopy, the integrated intensity of the light emanating from any one point in the specimen is unchanged as one moves a little away from focus, the light being merely redistributed in the final image. This explains why the conventional fluorescence microscope has a very poor resolving power in the direction of the optical axis.

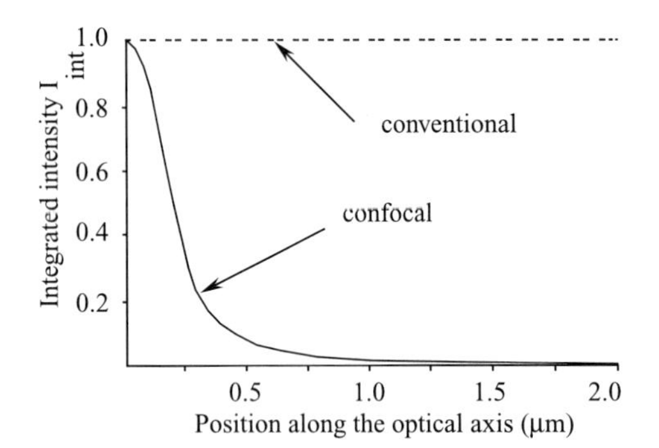

Figure 5.7 Depth discrimination in conventional (dashed line) and confocal (plain line) microscopy. These curves are obtained by summing the intensities in the *xz*PSF images of Figures 5.4 and 5.6, respectively, along horizontal lines (i.e., directions orthogonal to the *z* axis).

Conversely, the use of a circular pinhole aperture for confocal imaging causes the integrated light intensity to fall off sharply as one moves out of focus. It drops to half the original value at a distance of 0.7 λ from the focal plane for an objective of NA = 1.0 and an ideally small confocal aperture.

The following axial resolutions have been measured in different CLSMs:

- 730 nm (NA = 1.3, exc. 488 nm, em. 520 nm)
- 900 nm (NA = 1.25, exc. 514 nm, em. 590 nm)

It has been shown that the optical sectioning performance of a confocal microscope depends crucially upon several experimental variables:

- The wavelength of the illuminating light
- The Stokes' shift of the fluorochrome
- The numerical aperture and degree of aberration correction of the objective
- The size (and shape) of the confocal imaging aperture, and the nature of the specimen

➾ The Stokes' shift is defined as the difference between the emission wavelength and the excitation wavelength.

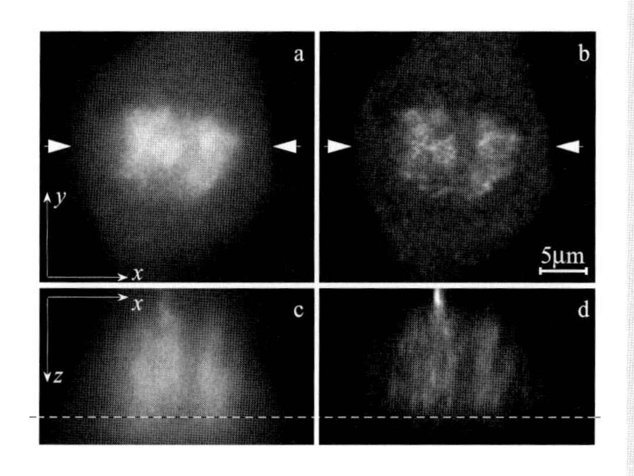

a and b: *xy* sections through a mitotic fibroblast, immunofluorescent staining of macroH2A histone proteins, the white arrowheads show the position of the *xz* scanning used to produce the images *c* and *d*

c and d: *xz* sections, the dashed line indicates the *z* position of the glass slide

The improvement of resolution provided by the confocal mode is very dramatic. One can also see that all fluorescent spots in image *d* appear elongated in the axial direction; this is the expected consequence of the difference of resolution between the *x* and *z* axes.

Figure 5.8 Comparison of conventional (left) and confocal (right) microscopy.

Thus, by its intrinsic optical properties, the confocal scanning fluorescence microscope rejects out-of-focus blur, and produces raw images that, to a first approximation, contain only in-focus information. This can be appreciated by comparing unprocessed confocal images with conventional fluorescence micrographs of similar specimens after computational removal of out-of-focus blur (Figure 5.8).

5.2.3.3 Light sources

The following gas lasers are widely used as light sources for confocal laser scanning microscopes (CLSM):

- Low-power (15 to 25 mW), air-cooled argon-ion lasers with lines at 488 and 514 nm
- Medium power (100 to 200 mW), water-cooled argon-ion lasers with lines at 458, 488, 514, and 529 nm
- High-power (3 to 5 W), tunable water-cooled argon-ion lasers with additional lines in the range 334 to 364 nm (used for UV excitation)
- Air-cooled argon–krypton lasers (about 50 mW) with lines at 488, 514, 568, and 647 nm

⇒ In most common biological applications, a laser power of 0.5 to 1 mW is found sufficient. Low-power lasers, such as argon-ion lasers with 15 to 25 mW power (in the multiline mode) are therefore convenient; however, they provide for only a limited number of lines. Lasers with higher power have to be used if a wide range of wavelengths is required.

⇒ Although an argon–krypton laser may appear as a very interesting source (many wavelengths), it is an expensive solution because of its short lifetime (3 years).

- Air-cooled helium–neon lasers with lines at 543 or 633 nm
- New optically pumped semiconductor laser sources have become available which provide rays in the visible domain

⇨ These solid-state sources will certainly overcome the classical gas lasers in the near future, because of their compactness and long lifetime.

5.2.4 The Two-Photon Microscope

Very recently a new type of fluorescence microscopy was introduced that offers most of the interesting features of confocal microscopy, while avoiding some ancillary drawbacks. It is called nonlinear fluorescence microscopy and is mostly referred to as "two-photon microscopy" (2PM).

⇨ Some people have already developed three-photon microscopy!

❑ *Advantages*

- High-resolution optical sectioning without the need of a confocal pinhole

⇨ Fluorescence emission can only be obtained within the focal plane of the objective lens.

- High penetration power through thick tissue specimens; it makes it possible to observe and image fluorescent labels through 250 μm thick tissue sections
- The use of tunable laser sources makes it possible to work with a wide variety of fluorophores

⇨ The excitation wavelength can be selected in the 700 to 1000 nm range.

- Well suited for fluorescence imaging of living cells.

⇨ The use of near infrared light to excite blue fluorophores (Hoescht 33342) avoids the damage generated by a conventional UV excitation.

- Reduced photobleaching when collecting series of optical sections

❑ *Disadvantages*

- Ultrafast tunable laser sources are quite expensive
- Requires special optics treated for working in the near infrared domain

5.2.4.1 Principle of nonlinear fluorescence

❑ *Nonlinear absorption*

The idea behind two-photon fluorescence microscopy can be summarized by this strange and simplistic formula:

$$1 \text{ photon} + 1 \text{ photon} = 1 \text{ photon}$$

On first view, this equation may appear absurd. However, if we consider the respective energy (in other words, the respective wavelength) of these photons, the equation becomes true under certain conditions. Let us consider that the two photons in the left side of the equation have the same wavelength, which is twice as long as that of the photon in the right side. In this case, our equation is true from an energy point of view. The equation can be rewritten:

$$\frac{\hbar c}{\lambda_1} + \frac{\hbar c}{\lambda_1} = \frac{\hbar c}{\lambda_2}$$

In other words, it is possible to excite a fluorochrome molecule with light at twice the required wavelength. Actually, this is only true under very particular conditions and can be explained by the simplified energy diagrams in Figure 5.9.

In conventional linear fluorescence, a single photon of right wavelength hits the molecule. The energy of this photon is absorbed by setting the molecule to its upper energy state (excited state). Then the molecule returns to its ground state in two steps, giving off a new photon of lesser energy. In the case of two-photon fluorescence, the excitation of the molecule to its upper energy state results from the quasi-simultaneous absorption of two photons of double wavelength.

⇨ The energy carried by a photon is inversely proportional to its wavelength:

$e = h\nu = h(c/\lambda)$; with e the energy, h the Planck's constant, ν the frequency, λ the wavelength, and c the velocity of light in vacuum.

⇨ With $\lambda_1 = 2\lambda_2$

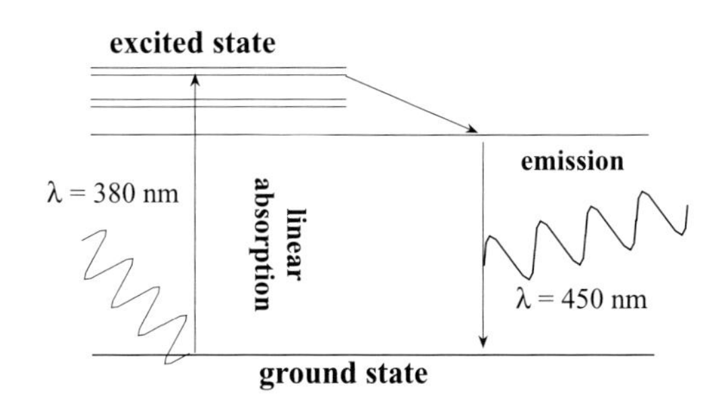

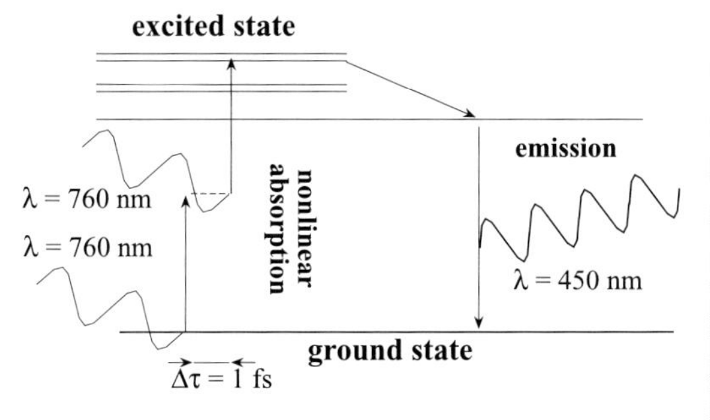

Top: single-photon fluorescence, a single photon (λ = 380 nm) is used to excite the molecule from its ground state to its upper energy state; then the molecule decays to the ground state in two different steps, giving off a new photon (λ = 450 nm).

Bottom: two-photon fluorescence, a single photon of half energy (λ = 760 nm) sets the molecule to intermediary excited state with a short lifetime. If a second photon arrives within a time-window $\Delta\tau$ of 1 fs (10^{-15} s) then the molecule is excited to its upper energy state, and the molecule returns to its ground state, giving off a new photon (λ = 450 nm).

Figure 5.9 Simplified tri-state energy diagrams of single-photon and two-photon fluorescences.

It is necessary that the second photon arrive very shortly after the first. In fact, the absorption of the first photon sets the molecule into a very brief "semi-excited" state. If no second photon hits the molecule in a lapse of one femtosecond (10^{-15} s), the latter returns to the ground state without emitting a photon. If a second photon hits the semi-excited molecule within the 1 fs window, it is absorbed and the molecule can reach its highest energy state and a photon is eventually emitted during the energy decay.

❑ *Sectioning power*

It is clear that very special conditions are necessary to achieve nonlinear two-photon excitation. As we have seen previously, two photons must hit the molecule in a very short time. In fact, if we consider a maximum delay of 10^{-15} s, this means the two photons must be very close in space as well. The maximum distance between them is:

$$d_{max} = c\Delta\tau = 3.10^{8}\ \text{ms}^{-1}\ 10^{-15}\ \text{s} = 0.3\ \mu\text{m}$$

➯ With c the velocity of light in vacuum and $\Delta\tau$ the time-window

Therefore a very high density of photons, that is, a very high light intensity, is necessary. Therefore, a powerful coherent light source must be used. Such a source is generally provided by mode-locked pulsed lasers, which can deliver very brief (80 to 200 fs) light bursts of high power (15 kW) at very high frequency (50 to 100 MHz).

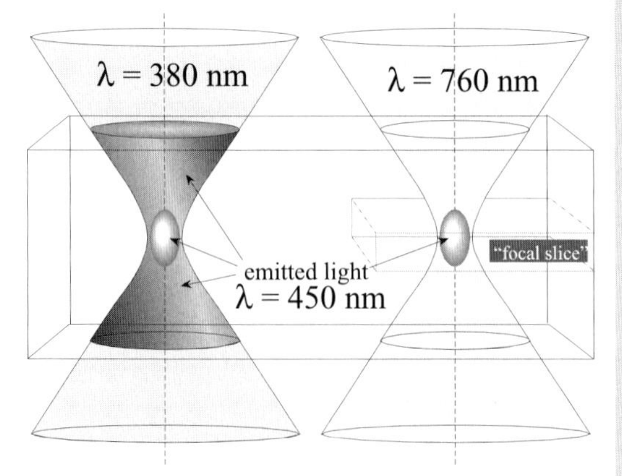

Left: single-photon fluorescence. Emitted light is produced with variable intensity inside the whole illumination double-cone.

Right: two-photon fluorescence. Emitted light is produced only in the focal plane of the objective lens.

Figure 5.10 Optical sectioning in two-photon fluorescence.

To obtain a high spatial and temporal density of photons, the pulsed light source is sharply focused through a large numerical aperture objective. As a matter of fact, the ideal conditions for two-photon absorption are reached at the focal point of the objective lens, and thus the fluorescence can only be obtained at this very point. Therefore, the respective optical sectioning properties of CLSM and of 2PM arise from two different processes. In the case of CLSM microscopy, fluorescence is excited everywhere through the aperture cone of the objective lens (Figure 5.10) and the pinhole aperture allows only light coming from the focal plane to reach the photodetector. Thus, the optical sectioning is the result of the imaging process.

In the case of two-photon microscopy, fluorescence is only produced at the level of the focal plane; the optical sectioning is a consequence of the illumination process.

➾ Those fluorochromes and tissues away from the focal volume of the objective lens do not suffer from photobleaching or phototoxicity, respectively.

5.2.4.2 Femtosecond laser sources

Very powerful coherent sources of infrared light are required to achieve an efficient two-photon excitation. This is provided by pulsed laser sources based actually on the combination of two lasers. A solid-state green laser of average power (between 5 and 10 W) is used as a coherent light source which pumps a second laser made of a titanium–sapphire crystal as dispersive medium enclosed in an adjustable length resonant cavity (Figure 5.11).

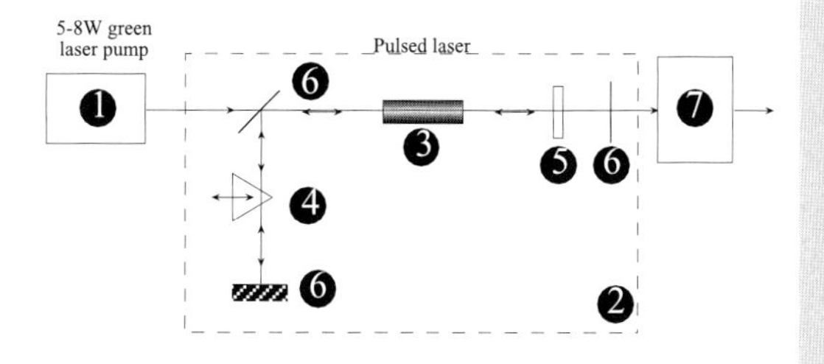

1. **Solid state green laser (5 to 10 W)**
2. **Resonant cavity**
3. **Titanium–sapphire crystal**
4. **Tuning prism**
5. **Mode-locking device (Kerr lens)**
6. **Mirrors**
7. **Pulse-width scaling unit**

Figure 5.11 Schematic diagram of a femtosecond pulsed laser.

The crystal is optically pumped by the green laser and generates light, the wavelength of which depends on the length of the resonant cavity. This produces a continuous coherent light with a wavelength longer than 700 nm and a rough output power of 1 W, which is not sufficient for 2PM. The wavelength can be tuned continuously in the range 700 to 1000 nm by changing the length of the cavity with a prism. In order to obtain a higher power the light path-length is modified with either a Kerr lens or a birefringent plate. This alteration of the cavity provokes a phenomenon of auto-oscillation called "mode-locking." When set in mode-locking, the laser delivers very brief (100 fs) bursts of light at very high frequency (100 MHz) and high power (10 kW).

⇨ 700 to 1000 nm wavelengths make it possible to work with conventional fluorophores, the excitation maximum of which is in the range of 350 to 500 nm.

⇨ It must be noted that although the instantaneous power of the light burst is high (10 kW), it is quite diluted over time. The average light dose received by the specimen is, in the end, rather low: during one second, the actual exposure time is 10^8 bursts of 10^{-13} s, that is, 10 μs. This is equivalent to illuminating the specimen during 1 s with a constant power of 10 mW.

5.3 LIGHT CAPTORS

5.3.1 Definition

The role of a light captor is to convert light intensity into a measurable electrical current. In fluorescence microscopy, the amount of light involved is small and sometimes only a few photons are actually available. Therefore, the captors must provide for a high sensitivity and an excellent signal vs. noise ratio.

⇨ Here, the signal vs. noise ratio (SNR) refers to light measurement. That is, the average number of photons emitted by the fluorophore vs. the random variation around this value over time. It must not be confused with labeling noise due to a loose specificity of the label.

There are two main categories of photodetectors: point captors and image captors. Both types are based on either photoelectric captors (vacuum tubes) or on solid-state captors.

5.3.1.1 Point captors

A point captor is made of a single photodetector. It is used mainly to measure the amount of light coming from a single point of the microscope field. The image of the whole microscope field must be reconstructed point by point. A scanning device, based either on oscillating mirrors or on a rotating Nipkow's disk, is used to successively select equally spaced points of the field. The detector measures these intensities, which are stored in the memory of a computer. Then the full image is reconstructed and displayed on a video monitor.

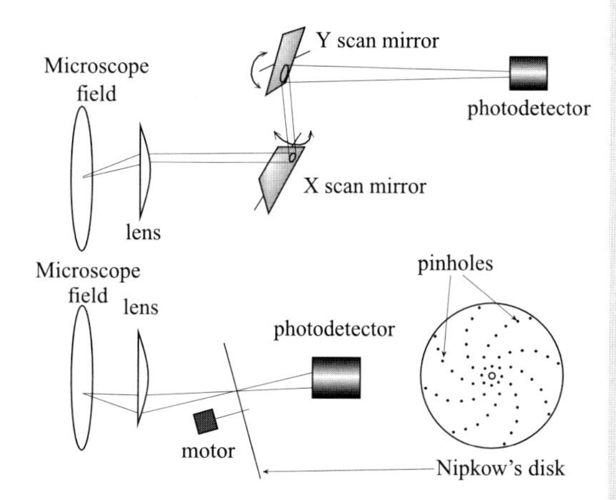

Top: scanning device based on two tilting mirrors

Bottom: scanning device based on a Nipkow's disk (small apertures arranged along intermingling Aristotle's spirals)

Figure 5.12 Point captors.

In general, these captors are used on imaging systems where sensitivity is the criterion of choice and acquisition speed is not important.

5.3.1.2 Image captors

An image captor is made of a regular array of point photodetectors. Thus, the image of a large area of the microscope field can be recorded at the same time. This category of captor is well suited for conventional microscopy.

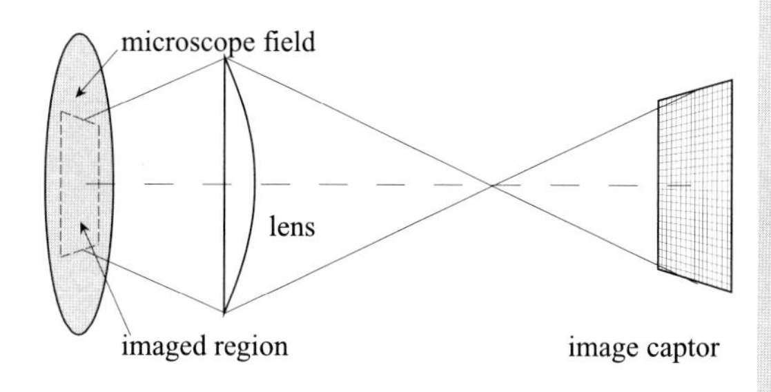

Figure 5.13 Image captors. A large area of the microscope field is projected on the surface of the detector made of an array of small photodetectors (CCD video camera).

5.3.2 Point Captors

5.3.2.1 Photomultiplier tubes

The photomultiplier tube or *PMT* is an electronic vacuum tube. The tube is made of a glass envelope coated with black paint over its surface except for a transparent entrance window at one end. Inside this tube, a series of electrodes can be found. The closest electrode to the entrance window is the photocathode (0 V).

⇨ *See* Figure 5.14.

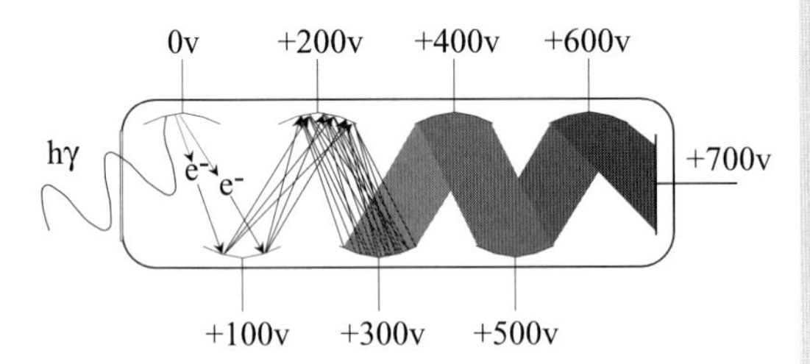

Figure 5.14 The photomultiplier tube. The photons ($h\nu$) enter the PMT through a transparent window and hit the photocathode (0 V). The photoelectrons (e^-) expelled by the collision are attracted by the next dynode of higher electrical potential.

When incoming photons hit the photocathode, some secondary electrons are emitted. These electrons are attracted to the next electrode (called a dynode), which is at a higher voltage. When the electrons collide with the first dynode, new secondary electrons are emitted in a growing number, which are, in turn, attracted by the second dynode. Therefore, between each dynode the number of secondary electrons is multiplied. In the end, a great number of electrons reach the anode and a current can be measured. It can be seen that the PMT is a very sensitive light detector; only a few photons are necessary to give a measurable current. It must be noted also that this current is directly proportional to the number of incoming photons and provides a wide dynamical range.

5.3.2.2 Photodiodes

The photodiode belongs to the category of solid-state photodetectors. A diode is composed of an *N* semiconductor crystal and a *P* semiconductor crystal glued together. The common face between the *N* and *P* crystals is called the junction. This *NP* junction behaves differently depending on the polarization of the voltage applied to the two electrodes of the diode (Figure 5.15). When a positive voltage is applied to the *N* crystal and a negative voltage to the *P* crystal, an electrical current can circulate through the *NP* junction. The positive electrode sucks the electron from the *N* crystal, while the negative electrode provides the *P* crystal with electrons. These electrons can jump through the junction to replace the electrons sucked by the positive electrode. This is called direct polarization.

➾ Semiconductor crystals are impure crystals grown from solutions of either silicon or germanium (tetravalent atoms), which are contaminated by a small proportion of either barium (trivalent atom) or arsenic (pentavalent atom). The addition of arsenic generates *N* semiconductor, which is an electron donor, while the addition of barium gives a *P* semiconductor, which is an electron acceptor.

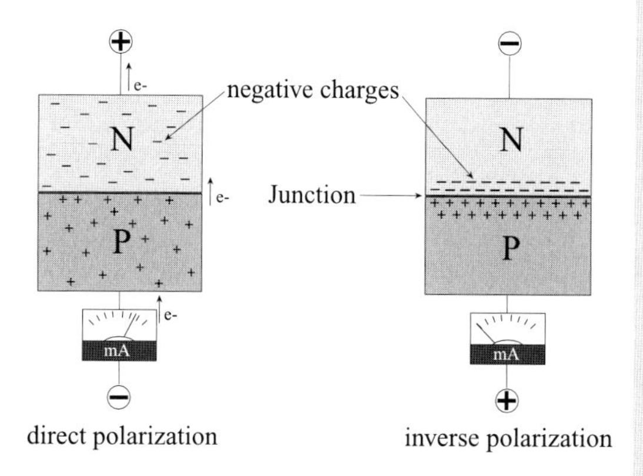

Left: direct polarization, the *N* crystal is connected to the positive electrode of the power supply and the *P* crystal is connected to the negative electrode. A current is established through the *NP* junction.

Right: inverse polarization, the *N* crystal is connected to the negative electrode and the *P* crystal to the positive electrode. No electrical current can circulate.

Figure 5.15 Principle of a solid-state diode.

Now, if the polarization is inverted by applying the positive voltage to the *P* crystal and the negative voltage to the *N* crystal, no electrical current can circulate. The negative charges of the *N* crystal are repelled by the negative voltage and accumulate along the *NP* junction, which becomes an energy barrier. This is called inverse polarization.

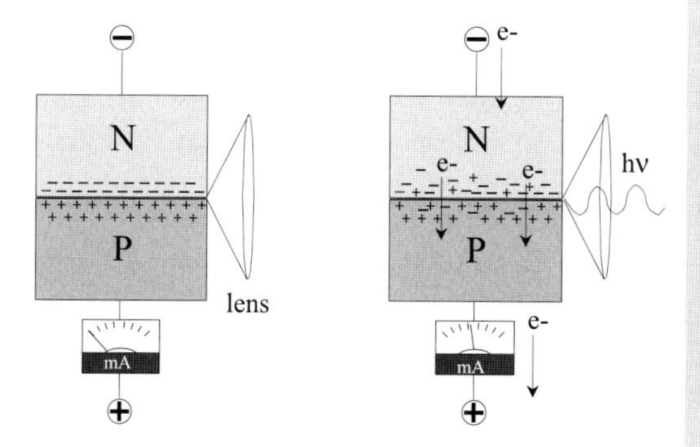

Figure 5.16 Principle of a photodiode. The photodiode is inversely polarized such that no electron can jump through the junction. An optical lens focuses light at the level of the junction. When photons reach the junction, they bring about enough energy to let some electrons traverse the junction. A brief electrical current is established.

The photodiode is a particular diode in which an optical lens focuses light at the level of the *NP* junction. The photodiode is inversely polarized such that no current can circulate (Figure 5.16). If light comes in through the lens, the photons that reach the *NP* junction momentarily disturb the electronic equilibrium at the level of the junction and let some electrons break the energy barrier. Thus, a current of electrons can circulate as long as photons hit the junction. This current is directly proportional to the number of incoming photons. Other solid-state photodetectors are derived from the principle of photodiodes. For example, phototransistors are the basic building elements of image captors. A special type of photodiode called *avalanche photodiode* has been developed which provides for a very high sensitivity. This photodetector is able to detect a single photon.

5.3.3 Image Captors

5.3.3.1 Architecture

The image captors are rectangular electronic chips whose main diagonal is in the range 1/3 inch to 1 inch. The sensitive surface is made of a large number (500,000 to 2,000,000) of small solid-state photodetectors. The photodetectors, which can be assimilated to pixels, are arranged on a regular lattice. Two main types of architecture are encountered. They are characterized by the read-out modality, that is, how the content of a single pixel is accessed: randomly or sequentially.

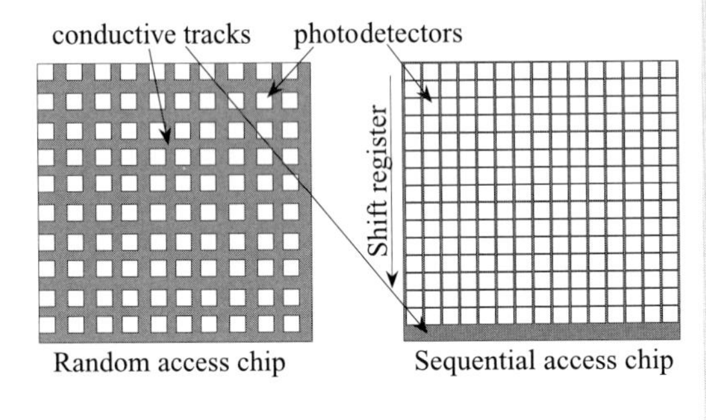

Left: random access architecture. Since all pixels must be accessed directly, a large area of the chip is used to drive currents between the pixels. The read-out time is short, but there are blind areas between pixels.

Right: sequential access. The pixels are organized in columns of shift register. The read-out time is long, but the blind areas are minimal.

Figure 5.17 Architecture of image captors.

5.3.3.2 Charge coupled device cameras

The charge coupled device cameras, known as CCD cameras, are image captors whose basic element is built around a solid-state photodetector associated with an on-chip small leakage capacitor. The electrons created at the level of the CCD element are stored in its capacitor. This architecture makes it possible to design captors capable of "on-chip integration." In situations where the photon number is very small, it is very important not to lose a single one. With on-chip integration, the probability of grabbing every single photon is enhanced by increasing the duration of the acquisition.

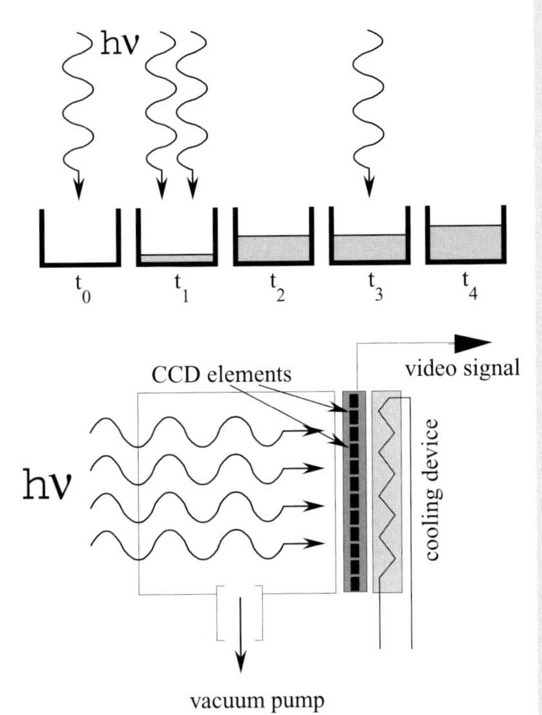

Top: on-chip integration. At time t_0, the shutter of the camera is open and the photons reach the chip. These are converted into electric charges at the level of the CCD elements. Charges are accumulated over time as a function of arriving photons. At time t_n, the total charge stored in a single element is read and the CCD is reset to null charge.

Bottom: cooling system. In order to improve *SNR*, the chip is set to a low temperature (0°C to –80°C) by a Peltier cooling device. For very low temperatures, it is necessary to avoid ice formation on the chip target. A vacuum pump prevents "wet" atmosphere from reaching the target.

Figure 5.18 Cooled CCD camera.

However, when on-chip integration times are greater than 1 s, the final signal may be contaminated by electronic noise. This noise, also referred to as dark current, is due to thermal agitation in electrical circuits. In a conductive material at room temperature, the friction between the molecules randomly releases electrons. These "noisy" electrons contaminate the charge of the CCD element. Thus, the charge in a CCD is proportional to the number of photons (significant signal) plus a random amount due to thermal electrons (noise). In order to improve *SNR*, it is necessary to reduce the occurrence of thermal electrons. This can be achieved by reducing thermal agitation — in other words, by lowering the temperature of the chip. A Peltier cooling device is stuck to the back of the CCD chip to lower its temperature. The Peltier device is a heat pump; it pumps the calories from the chip and dissipates them by convection. The heat convection can be improved by adding a metal heat-sink and an electric fan.

5.3.4 Light Intensifiers

Cooled CCD cameras with long integration time and read-out are not suited for fluorescence applications where speed is critical. This is the case for kinetics experiments where it might be necessary to record images at video rates (25 frames/s) or even faster. Light intensifiers offer a satisfactory solution. These are optical or photoelectronic devices that amplify incoming light and are placed in front of an image captor. They behave like photon multipliers: the number of outcoming photons is greater and proportional to the number of incoming photons. Two principal types of light intensifiers can be described:

- Solid dye amplifiers
- Photoelectronic devices

5.3.4.1 Solid dye amplifiers

Solid dye amplifiers are based on the properties of some crystals which have to act as light dispersive media. These crystals are of the same type as those used in solid laser. The light intensifier is composed of a large number of small crystal bars of cylindrical shape. These small bars are called microchannels. They are regularly packed together and define an entry face and an output face. The entry face is the sensitive extremity of the device. The output face is associated with an image captor, which collects the intensified image.

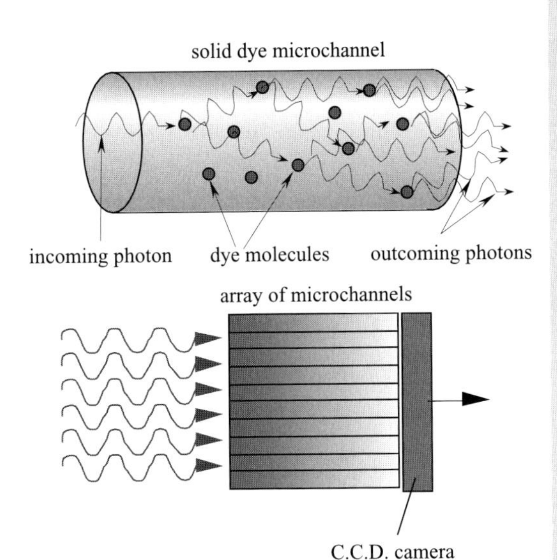

Top: principle of light amplification by a microchannel.

Bottom: architecture of a solid-dye light intensifier. The microchannels are assembled to form an array of light multipliers. In order to preserve image resolution, the number of microchannels per camera pixels must be greater than or equal to one.

Figure 5.19 Solid dye light intensifier.

When a photon enters the sensitive end of a microchannel, it collides with the dye molecules. These emit other photons that will in turn hit other molecules. Thus, when a photon enters the microchannel it starts a chain reaction within the crystal that results in a multiplication of photons. The trajectories of the created photons are constrained by the cylindrical shape of the bar, and are driven toward the faces of the microchannel. Those photons coming out the output face are collected by the image captor (CCD camera at video rate).

5.3.4.2 Photoelectronic intensifiers

Photoelectronic light intensifiers employ a photoelectronic valve and a video captor. The valve is a specially designed vacuum tube. A large photocathode covers the entry face of the tube. When a photon hits this electrode, it brings about energy enough to expel a secondary electron from the surface of the electrode. This electron passes through a hollow anode. The anode has the shape of truncated cone and creates a gradient of the electrical field. The consequence of such a gradient is to accelerate the electron. In the end, the electron hits a phosphor screen with high energy. A bright spot appears at the location of the collision, and this can be imaged by a CCD camera.

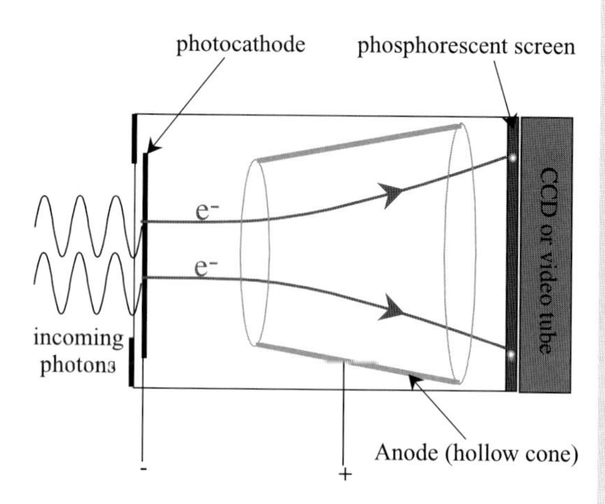

Figure 5.20 Electronic light intensifier. The incoming photons hit the photocathode. Secondary electrons are emitted and accelerated through a hollow anode. These photons hit the phosphorescent screen where bright spots appear. A CCD camera is placed behind the screen to record the intensified image.

Chapter 6

3-D Visualization and Image Processing

Yves Usson

CONTENTS

6.1 VISUALIZATION OF 3-D DATA SETS

6.1.1 Exploration Tools

The exploration tools make it possible to browse quickly through large stacks of serial sections. There are two basic tools: the gallery display and the orthogonal projections.

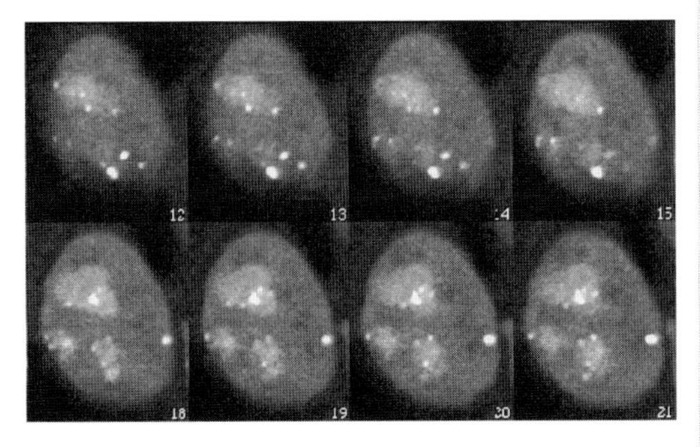

Figure 6.1 Gallery display of a series of optical sections through a nucleus of a HeLa cell stained with IP. The FITC labeling of HSF1 (heat shock factors) can be seen as small bright spots. Only eight sections out of 40 are shown.

In the second case, the user can explore the data set through a friendly interface. The latter displays three orthogonal section planes cut through the whole data set. The main section is an *xy* section, which corresponds directly to an optical section. The two other sections are reconstructed views along planes that are perpendicular to the optical section. One is built along a single row of the *xy* section and extends along the *z* axis; it is called an *xz* section.

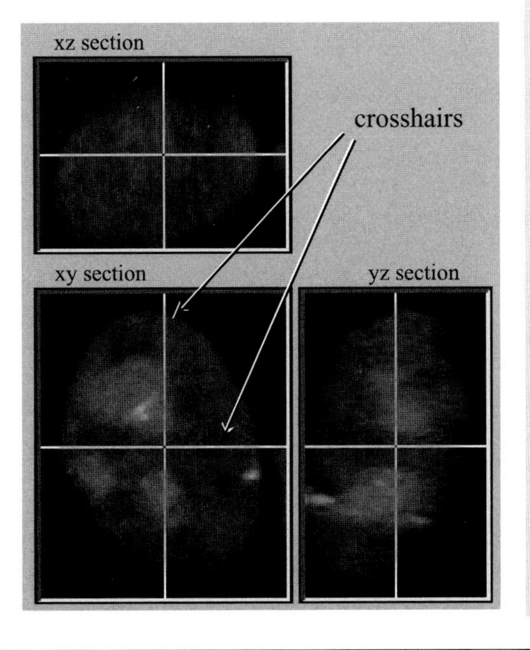

Figure 6.2 Orthogonal sections of the same data set as in Figure 6.1. The *xy* section (bottom-left) corresponds to an actual optical section as recorded with the confocal microscope. The *xz* and *yz* are reconstructed sections. These are built by stacking the fluorescence information along the lines defined by the crosshairs and as a function of depth. The user can interactively set the crosshairs at any position in order to explore the volume.

0-8493-0817-8/01/$0.00+$1.50

The second is built along a single column of the xy section and extends along the z axis; it is called a yz section. These three orthogonal sections are displayed on the computer screen in a manner similar to architecture blueprints (Figure 6.2). The level of the sections can be changed interactively by dragging crosshairs on top of the images.

6.1.2 Visualization Tools

The visualization tools are based on 3-D reconstruction techniques. The basic principle is to create either a still or animated synthetic image of the volume. Different visual attributes can be selected in order to render specific information. The two main attributes are:

- The shape of the observed objects (cells, nuclei, etc.); it implies a "surface rendering" approach
- The content of the objects; it implies a "transparency rendering" of volumes

➪ In this case, the reconstructed objects appear as solid and opaque structures.

➪ In this case, the reconstructed objects are translucent.

6.1.2.1 Surface rendering

The first step of surface rendering is the segmentation of the 3-D volume.

a. Segmentation

The segmentation consists in identifying the voxels that belong to an object. In fluorescence, this can be easily achieved by a simple intensity thresholding. All those voxels with a greater value than the intensity threshold are considered object voxels; the others are background voxels.

➪ A voxel is short for "Volume Picture Element." It is the smallest element of a 3-D discrete image. It is characterized by its x,y,z position, its x,y,z size, and its intensity value.

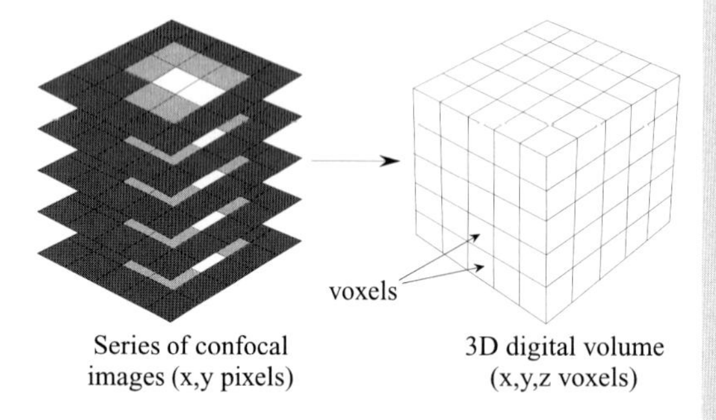

Figure 6.3 Digital representation of a 3-D data set. The set is made of a series of confocal sections stored as digital images (x,y pixels). These images are "piled" in the memory of the computer to form a 3-D digital volume made of voxels. The z dimension of a voxel corresponds to the distance between two consecutive sections.

The surface extraction is obtained using a basic neighborhood rule: if a voxel belongs to the object and if at least one of its six direct neighbor voxels (above, beneath, left, right, front, and back) belongs to the background, then the voxel belongs to the surface.

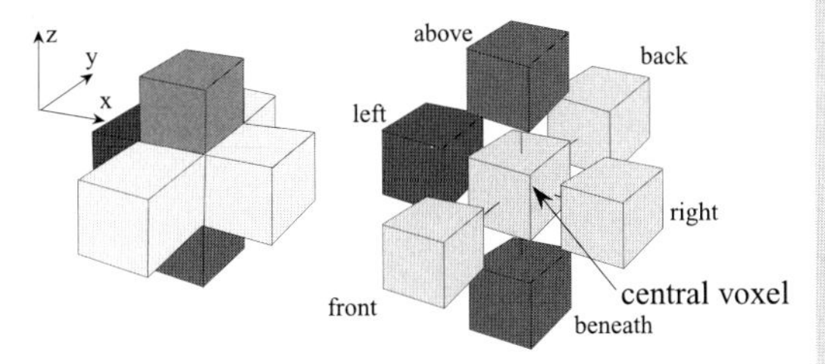

Figure 6.4 Segmentation rule: the central voxel is surrounded by six other voxels that share a common face with it. The voxels are shown as cubes whose brightness is proportional to fluorescence intensity. The central voxel belongs to the surface of the object because its intensity and the intensities of three of its neighbors (front, back, and right) are greater than the selected threshold.

b. Depth map and Z-buffer

An easy and fast representation of the 3-D structure is obtained by depth encoding in a Z-buffer. The Z-buffer is a two-dimensional array (*x,y* coordinates) set parallel to the video display screen. The method consists in piling up the surface voxels according to their distance from the viewing screen. The Z-buffer is first initialized to infinity. Then the *x*, *y*, and *z* coordinates of each surface voxel are taken into account as follows. The distance from the screen of the current voxel is compared to the distance stored in the Z-buffer for the same *x*, *y* coordinates.

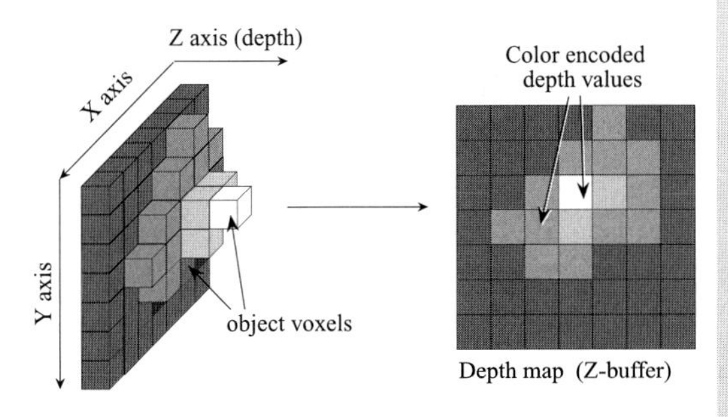

Figure 6.5 Depth map (Z-buffer): it may be compared to piling up the voxels as a function of the distance to the viewing point.

If this distance is smaller than the one already stored in the Z-buffer, then the distance of the current voxel replaces the value in the Z-buffer. At the end of the process, the distances are converted to intensity values or false colors, and a depth map (or topographical map) can be viewed on the screen.

⇨ Although it contains very relevant information, it may be difficult to grasp. Operators must train themselves to read such a graphical representation.

c. Illumination and shading

The *Z*-buffer is in fact a topological representation of the reconstructed 3-D structure. To obtain a "more realistic" display it is possible to apply a simple shading algorithm directly on the content of the *Z*-buffer. Diffuse reflection shading is the simplest shading model. Despite a low level of sophistication, this models offers a fast and efficient rendering of confocal 3-D sets. Let us consider a flat surface. The normal vector *N* to the surface is a vector perpendicular to the surface plane. The amount of diffuse light reflected by this surface is proportional to the cosine of the angle (α) between vector *N* and the direction of the incident light. In such a simplified model, the direction of the incident light is supposed to be parallel to the viewing direction.

➪ "Realistic" must be placed in quotes. In fact, we build a solid model of the observed structure and display it with arbitrary attributes that we may expect to be more explicit to the observer.

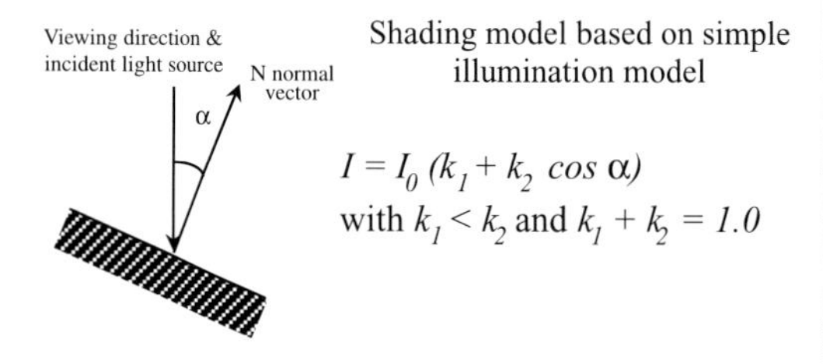

Figure 6.6 Shading model. The model mimics light diffusion by an oblique planar surface. The amount of light *I* viewed by the observer is proportional to the cosine of the angle *a* between the incident light I_0 and the normal *N* to the plane. k_1 (proportion of ambient light) and k_2 (proportion of diffuse light) are weighing coefficients.

The normal vector of each visible voxel stored in the *Z*-buffer can be obtained by calculating the local depth gradient. This gradient is defined by the difference of altitude between the current voxel and its four direct neighbors. Then the quantity of diffuse reflection can be calculated for the current voxel using the cosine of the angle between the normal to the voxel and the direction of the incident light (which is generally assumed to be parallel to the viewing direction).

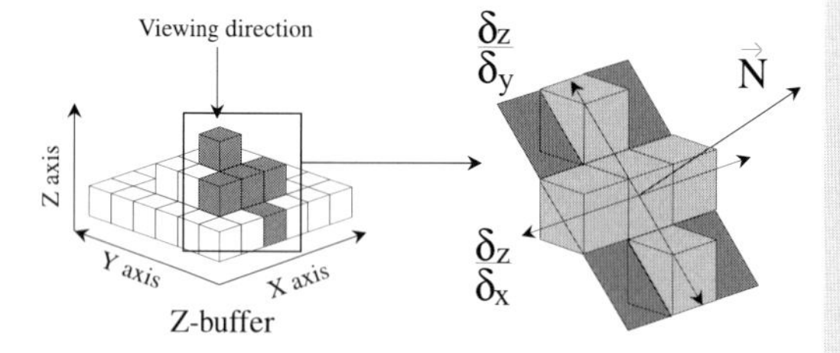

Figure 6.7 Estimation of the normal vector to the surface in a *Z*-buffer. The direction of the vector is derived from the local *z* gradient along the *x* direction ($\delta z/\delta x$) and the local gradient along the *y* direction ($\delta z/\delta y$).

When these simple algorithms are applied to confocal data series, we obtain a final reconstruction of segmented objects. For example (Figure 6.8), after reconstruction of the data shown in Figure 6.1, the nucleus of the HeLa cell appears as a solid almond with a shiny, embossed surface. The HSF1 signals appear as small shiny clumps.

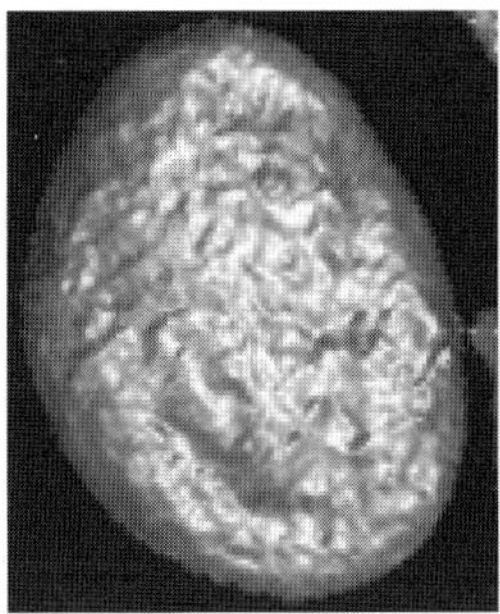
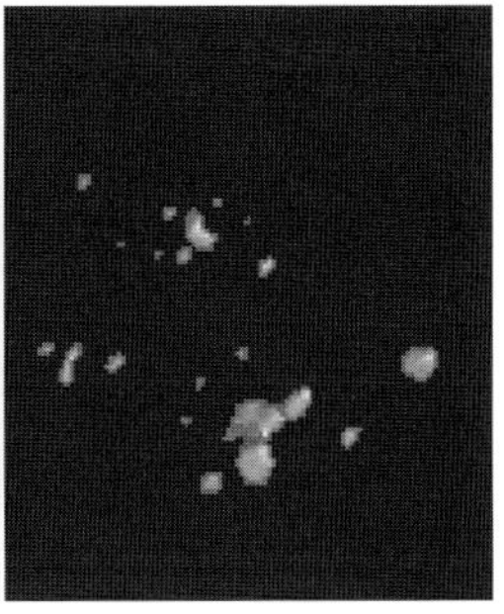

Figure 6.8 3-D reconstruction of a HeLa cell nucleus (left) and of HSF1 signals (right). The surface rendering based on a simple illumination model generates images with a shiny, metallic look.

6.1.2.2 Transparency volume rendering

In contrast to the surface model, where only the surface voxels are considered, the volume model is intended to show all the information contained in the 3-D cube. This display mode might be called "see through" or transparent viewing.

a. Extended focus

The extended focus consists in simulating a microscope equipped with a high resolution objective (high *NA*) but with an extended depth of field (low *NA*). This is obtained by building up an image made of the sharpest structures in the different focal planes. For example, in the case of transmitted light, we calculate the local sharpness indices for a given pixel (with coordinates x,y) throughout the different focal planes (z). Then we retain the intensity value in the focal plane that had the highest local sharpness index. In the case of fluorescence, we retain the brightest (maximum) intensity at coordinates x,y throughout the different focal planes (z).

➾ Actually, only a computer can do this. Such an ideal microscope based on optical parts is still a wish.

➾ Local sharpness can be defined as the local contrast in an image, that is, a measure of the second derivative of light intensity. This calculation is a trivial task for a computer.

b. Ray casting

The method for representing the volume is called ray casting. The viewing screen is divided into small, elementary points (pixels). Then we consider that each pixel of the viewing screen receives a ray of light originating from infinity and which traversed the 3-D data volume. The intensity of a given ray is the result of the information encountered along its path. In other words, the light intensities of all the traversed voxels are accumulated and contribute to the brightness of the pixel displayed on the screen.

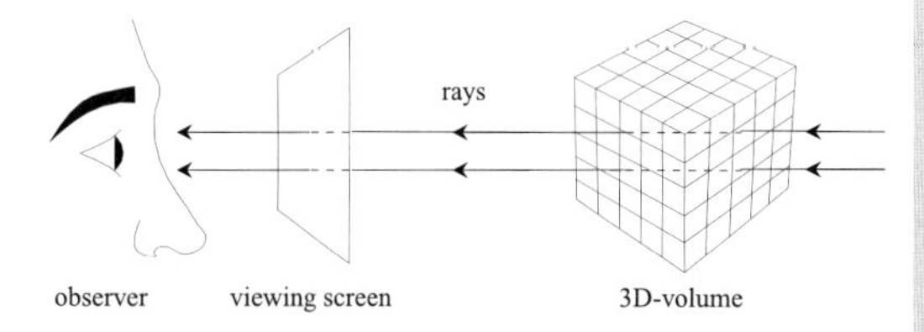

Figure 6.9 Principle of ray casting. Virtual rays, coming from "far away," come across the 3-D volume and intercept the viewing screen. The intensity displayed at the interception point is the "history" of the ray.

c. Transparency models

If the whole information collected along the ray was simply displayed on the screen, this would result in a blurred image since all the voxels have the same weight regardless of their distance from the screen. The representation can be greatly improved by introducing a transparency rule that preserves the sharpness of the details as well as respecting their depth hierarchy. As the ray encounters a voxel (v) its intensity value (I_v) is weighted by a transparency coefficient ($\alpha(I_v)$) and is added to the previous value of the ray (R_{v-1}).

⇒ The advantage of using a confocal microscope is lost.

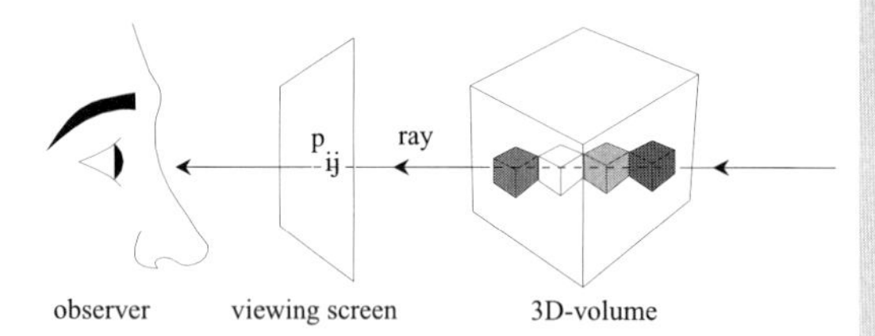

Figure 6.10 Transparency rule. Each traversed voxel emits a given amount of light. This amount contributes to the final intensity of the ray but also partially hides the light collected earlier along the ray path.

$$R_v = \alpha(I_v)I_v + [1 - \alpha(I_v)]R_{v-1}$$

⇒ v is the currently traversed voxel; I_v is the intensity of this voxel; R_{v-1} is the value carried by the ray up to the current voxel; and R_v is the new value.

The coefficient value depends on the voxel intensity; the higher the intensity of the pixel, the smaller its transparency coefficient. Different translucency effects can be obtained using different functions for this coefficient:

$\alpha(I_v) = kI_v$, linear law,

$\alpha(I_v) = k\sqrt{I_v}$, square root law, or

$\alpha(I_v) = k\log(\varepsilon + I_v)$, logarithmic law.

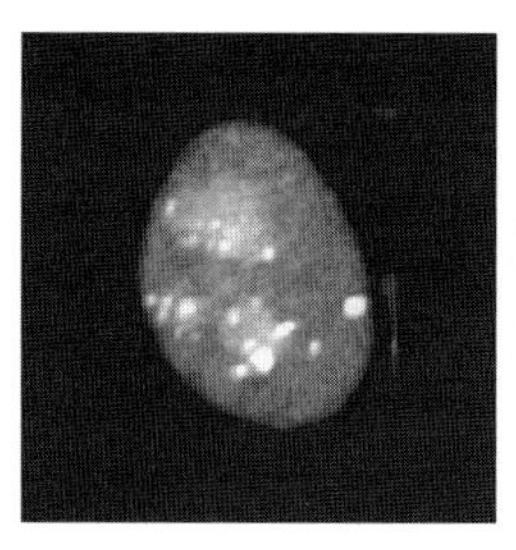
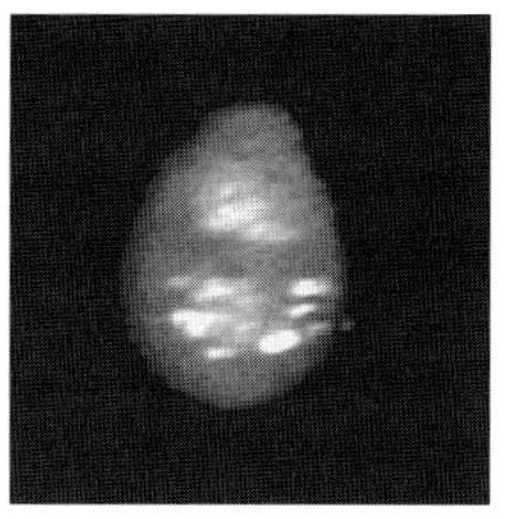

Left: face view.
Right: side view (rotation of 70° about the y axis).

Figure 6.11 Ray casting reconstruction of the HeLa cell nucleus (*see* Figure 6.1).

d. Depth cue — Stereovision

The previous methods result in flat images displayed on a flat screen. As for classical photography, depth cue is lost in the display process. However, it is possible to retrieve this depth feeling by using the principle of stereovision. The depth cue is, in fact, the result of the processing of the images sent to the brain by both eyes. The image seen by the left eye is slightly different from the image seen by the right eye. Our brain interprets these small differences to reconstruct the spatial relationships between objects in the 3-D scene.

Therefore, depth cue can be simulated by taking advantage of the stereovision principle. The basic idea is to calculate two separate 3-D reconstructions of the same scene, as if these were seen with the left eye and the right eye, respectively. In the end, these two images are made available to the observer in such a way that each eye only sees the image it is supposed to see. This can achieved by various techniques: polarized slide projectors, stereo viewers, color anaglyphs, or video rate switched images.

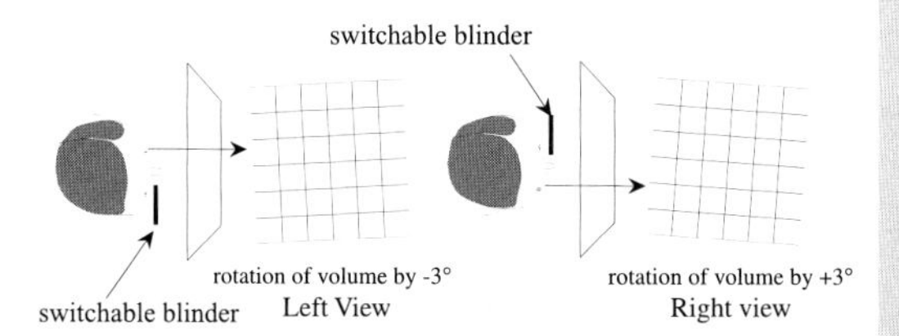

Figure 6.12 Depth cue and stereovision. Alternative display of the two images at high rate. The left eye is blindfolded while the right image is displayed; then the right eye is blindfolded while the left image is displayed.

- Anaglyphs: the left image is colored in red and superimposed on the right image colored in green; the result is observed with bicolor goggles (red glass for the left eye and green glass for the right eye).

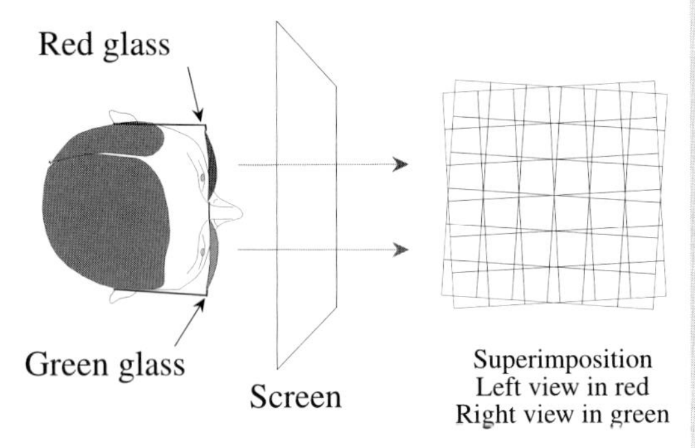

Figure 6.13 Color anaglyphs. The images of the stereo-pair are superimposed with different colors, left in red and right in green. The observer wears goggles with a red glass on the left side and a green one on the right side.

- Switched display: The two images are alternatively displayed on a computer screen at high rate. The observer wears special goggles made of liquid crystal glasses. These glasses act as switchable blinders that are driven in synchronicity with the image display.

⇨ The switching frequency must be greater than 50 Hz in order to avoid flickering.

- Polarization: the two images are projected on a screen through cross-polarized filters. The result can be observed with special goggles with cross-polarized glasses.

⇨ A horizontally polarized filter is placed in front of the lens of the left slide projector and a vertically polarized filter is placed in front of the right slide projector. The observer wears goggles with a horizontally polarized glass on the left side and a vertically polarized glass on the right side.

6.2 PROCESSING OF 3-D DATA SETS

6.2.1 Introduction

Fluorescence confocal microscopy provides large data sets that require powerful computer software to render and analyze. *CLSM* must be thought of like other measurement tools. Although *CLSM* is a very expensive and sophisticated apparatus, the images produced are not free of defects. These are due to the limitations of the imaging process and are listed below:

- Poor signal vs. noise ratio

⇨ The detection pinhole drastically limits the amount of light collected by the photodetector. Only light coming from the focal plane is measured, and it corresponds to a very small quantity.

- Absorption of light within thick specimens

⇨ The slight variations of the refraction indices within the cells and tissues create light scattering.

- Anisotropy of the spatial resolution
- Geometrical distortions

⇨ The axial resolution is two- to threefold worse than the lateral resolution.
⇨ A loose match between the refraction indices of the optical parts (specimen, embedding medium, glass coverslip, etc.)

Fortunately, these limitations can be modeled and compensated for by using the power of digital image processing.

6.2.2 Noise Preprocessing

The images obtained with a fluorescence microscope have a bad signal vs. noise ratio (SNR). The latter is even worse for confocal microscopy. This is due to the low level of light and to the quantum nature of photons. If we observe a point of light source with a given brightness during a time lapse t, the probability of gathering a given number of photons is distributed like Poisson's law. This means that for a region of the sample the measured brightness is equal to an average value μ with a standard deviation σ equal to the square root of μ. The consequence is that the SNR is always better in bright regions than in dark regions.

⇨ Here, by *signal* we mean "light intensity," and by *noise* we mean "random variation of light intensity." It must not be confused with the efficiency of FISH or immunolabeling.

⇨ SNR can be calculated as follows:

$$\mathrm{SNR} = 10 \log_{10}(\mu/(3\sigma))$$

and is expressed in decibel (*dB*).

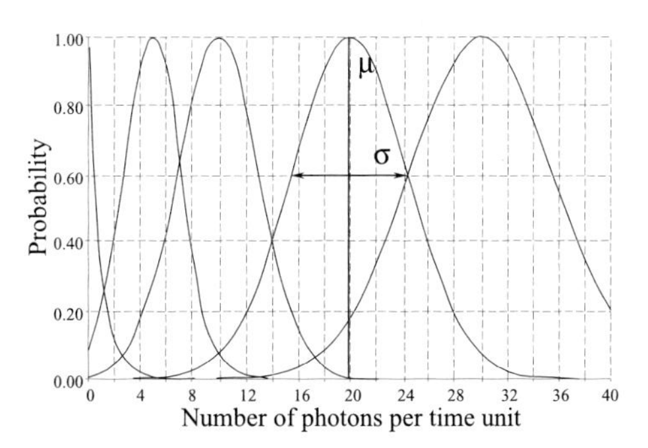

Figure 6.14 Statistical distributions of photon noise. Distributions for light sources emitting an average number of 1, 5, 10, 20, and 30 photons per time unit. The standard deviation σ is equal to the square root of the average intensity μ.

Different approaches can be used to improve the SNR. One way is to increase the acquisition time. Thus, the number of collected photons is also increased. Another method is to record k frames of the microscopic field and then to calculate an image by averaging pixel by pixel the k recorded images. This way the SNR is improved as a function of $\sqrt{k}$.

An example of such processing is shown Figure 6.15. Three images of a confocal section through a nucleus of a human myocyte (TOPRO III labeling) were obtained after different frame averaging. On the left, the raw image (that is, with no averaging) is very noisy. The image has a "salt and pepper" look, which characterizes photon noise. The details of chromatin texture are hardly visible. As can be seen in the two other images, the SNR is dramatically improved by averaging multiple frames. With 32 frames, we obtained a well-defined image.

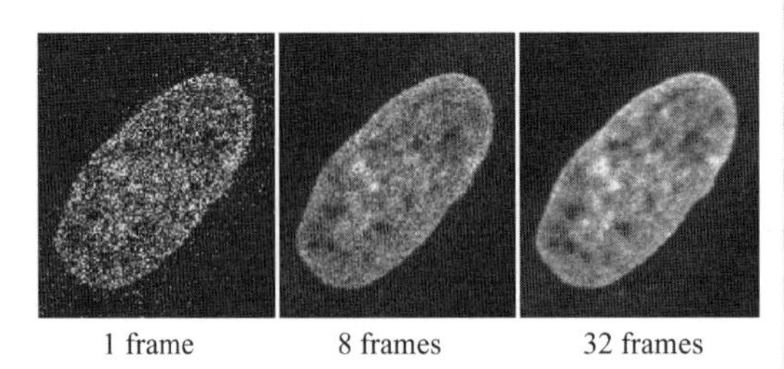

Left: image of a myocyte nucleus obtained after recording a single confocal frame

Center: averaging of 8 frames

Right: averaging of 32 frames

Figure 6.15 Enhancement of SNR by frame averaging.

However, this technique requires a long exposure time and the fluorochromes are very sensitive to fading (*see* Figure 6.16). In particular, in CLSM, when we collect the emitted fluorescence from the focal plane, the planes above and under are excited as well and suffer from photobleaching. Thus, a compromise must be found between expected SNR improvement and acceptable photodegradation.

⇒ Observing a fluorescent specimen impairs its fluorescence! Therefore, it is very important to minimize the exposure time as much as possible.

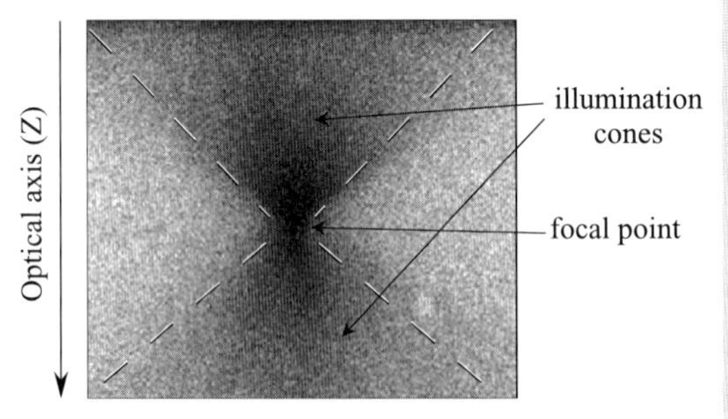

Figure 6.16 Fluorescence bleaching. *xz* section through a gel stained with FITC. A laser source was focused for 2 min by the objective lens of the microscope. The fluorescence has completely vanished at the focal point and diffuse photodegradation is also observed within the illumination cones above and under the focal point.

An alternative is to use image restoration techniques. Here, the SNR improvement is achieved by numerical methods relying on a good modeling of the imaging process and the nature and amount of noise involved. Different types of digital filters can be used:

- Linear filters (Gaussian)
- Rank order filters (median)
- More sophisticated techniques combining linear and rank order filters

These filters present different advantages and drawbacks. In general, the simplest filters such as Gaussian or median filters provide for a very efficient noise rejection. However, they tend to impair the image resolution. In other words, they introduce more blur than is already present, and the edges and fine details might not be preserved.

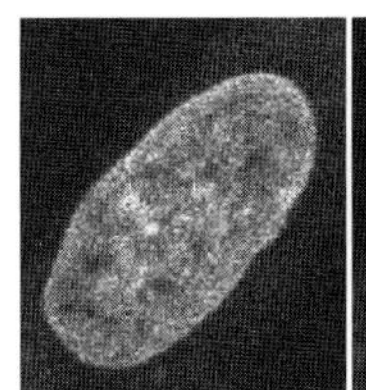

8 frames

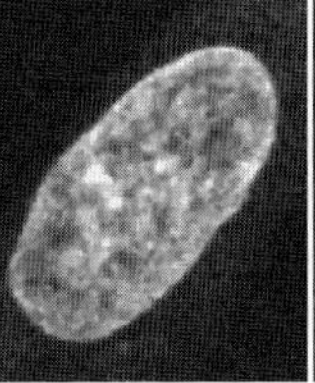

Median filter

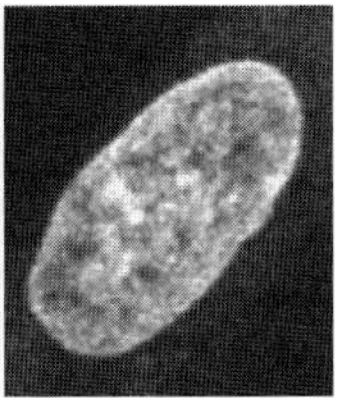

Gaussian filter

Left: confocal section obtained by averaging 8 frames

Center: 8 frames and median filtering

Right: 8 frames and Gaussian filtering

Figure 6.17 Digital noise filtering.

Digital adaptive filtering makes it possible to efficiently smooth out noise while preserving edges and fine details. As a matter of fact, the adaptive approach requires two different stages:

- A first stage where local statistics (variance or gradient) are calculated for each pixel
- The filtering stage itself where the type of filter and the degree of filtering are modulated as a function of the local statistics obtained in the first stage

⇨ An adaptive filter may be considered an "intelligent" filter. Instead of brutally applying the same process on all the pixels or voxels of the volume, the amount and the nature of noise is first assessed locally to select an adequate filtering method and strength.

6.2.3 Compensation of Light Absorption

Although the biological samples are, in general, transparent, they tend to absorb a slight amount of light. Furthermore, when light comes through a membrane, a nucleus, or subcellular organites, a given amount is diffracted. In specimens thicker than 5 μm, the absorption phenomenon is no longer negligible. First, the excitation light is gradually attenuated as a function of depth on the way in. Second, the emitted light is also attenuated on the way out. This results in a significant drop in the signal intensity as a function of *Z*.

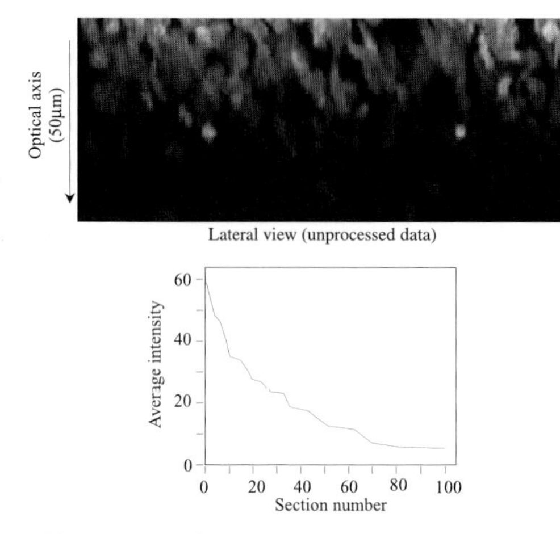

Top: side view of the 3-D reconstruction of a tissue section (myocardium); nuclei are labeled with Feulgen stain. No nucleus can be seen after a depth 25 µm.

Bottom: plot of the average fluorescence intensity measured as a function of depth within the tissue. The average intensity decreases.

Figure 6.18 Light absorption as a function of depth.

The curve of attenuation as a function of depth can be modeled by a log–logistic function. Therefore, it is quite easy to design a scheme to compensate for light absorption. It can be done either "in-line" during the acquisition of the serial sections or "off-line" by subsequent digital processing. The in-line correction requires that the gain of the photomultiplier be monitored during the acquisition. The gain must be adjusted automatically between each section. The off-line correction consists in multiplying the gray levels of the pixels by a coefficient or gain varying as a function of depth.

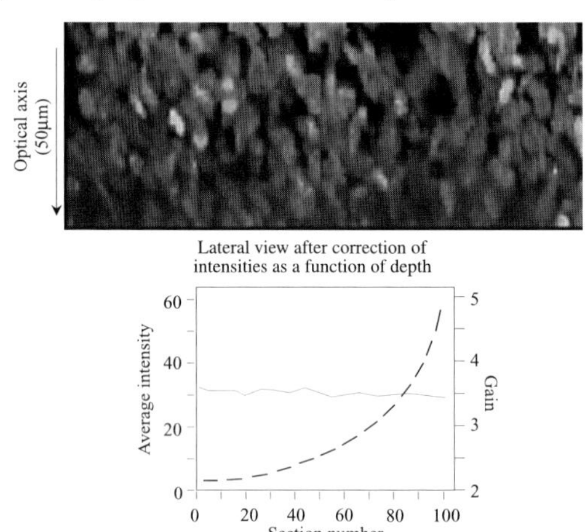

Top: side view of the 3-D reconstruction of the same data as in Figure 6.18 after correction.

Bottom: plots of the correction gain (dashed line, right scale) and the average fluorescence intensity measured after correction (left scale)

Figure 6.19 Correction of light absorption as a function of depth.

An example of off-line gain correction is shown in Figure 6.19. We can appreciate the good equalization of the intensity levels throughout a 50-µm-thick sample. However, it must be noted that the SNR is not evenly distributed as a function of depth. As a matter of fact, it is worse in the deepest sections than in the shallowest.

➾ The gain correction for the deepest section amplifies signal as well as noise! In fact, an adaptive noise filtering must be coupled to the gain correction.

6.2.4 Axial Resolution Enhancement

As we saw in Chapter 5, the axial resolution is nearly threefold worse than the lateral resolution. As a consequence, a voxel is not a cube but rather a brick-shaped element with the longest dimension parallel to the optical axis. Therefore, all measurements must take this anisotropy into account. A way to reduce it consists in interpolating virtual sections between the optical sections. This is made possible by digital resampling of data along the *Z* axis. Resampling can be performed using different approaches, such as linear or cubic spline interpolation. However, prior to interpolation, it is necessary to improve the axial resolution. This is obtained by digital deconvolution.

6.2.4.1 Digital deconvolution

As we have seen previously (Chapter 5, Section 5.2.2.2), the imaging process by a lens can be modeled by a convolution operation. The convolution transforms the distribution of a point light source in the object plane into a diffraction spot (Airy disk) in the image plane. It is possible to reverse this transformation by using numerical techniques. Thus we can expect to restore, at least partially, the original information.

⇒ This reverse transform is called *deconvolution*, since it aims at canceling the effect of convolution by the objective lens.

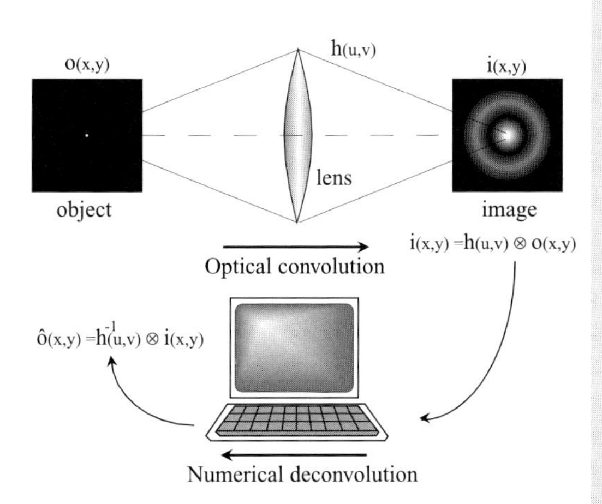

Top: the lens performs a convolution of the object (*o*) to form an image (*i*). The lens is characterized by its PSF or convolution kernel (*h*).

Bottom: numerical deconvolution is performed with a computer. The estimate ($\hat{o}$) of the object is obtained by numerically "convolving" the image (*i*) by the reverse PSF (h^{-1}).

Figure 6.20 Definition of image deconvolution.

The restoration of the object *o* from the image *i* can be reduced to an operation of linear inverse filtering expressed as $o = h^{-1} \otimes i$. Unfortunately, in the real world it is very difficult to obtain h^{-1} and, furthermore, the image is always corrupted by photon noise.

⇒ Inverse linear filtering is the most direct way of "deconvolving" a data set. It is based on a perfect knowledge of the reverse PSF and requires a very good SNR to be applied successfully.

Iterative constrained deconvolution has been developed to overcome these difficulties. In such an approach, we look for a fair estimate ($\hat{o}$) of the object. This estimate is then refined by slight corrections. Between each iteration, we assess the likelihood of the estimate by submitting it to the same blur as obtained with the objective lens. Then we compare, pixel by pixel, the blurred estimate ($h \otimes \hat{o}$) with the actual fluorescence image (i). In the hypothesis where the estimate is perfect, there should not be a difference. If this is the case, we may stop the restoration process. Otherwise, this means that the estimate is not a good one and we must correct it. For this purpose, we can use the pixelwise differences between $h \otimes \hat{o}$ and i to correct the estimate.

Many algorithms based on the same principle have been developed by different authors. They differ mainly in the way the error of estimation is measured, how the object estimate is corrected, and how it deals with noise. In general, a satisfactory restoration can be obtained with a small number of iterations (between 5 and 10).

➫ Iterative constrained deconvolution can be compared to a painter drawing a lady's portrait. In a first step, he draws a sketch with the global shape and the main lines of the visage. This may be considered the first estimate of the object. Then he compares the raw sketch with the visage of the lady and starts to refine his drawing by slight retouches. This is equivalent to the correction stage of deconvolution (the blurred estimate is compared to the actual image). The painter repeats (or iterates) the retouching step until he finds the resemblance of his drawing satisfying.

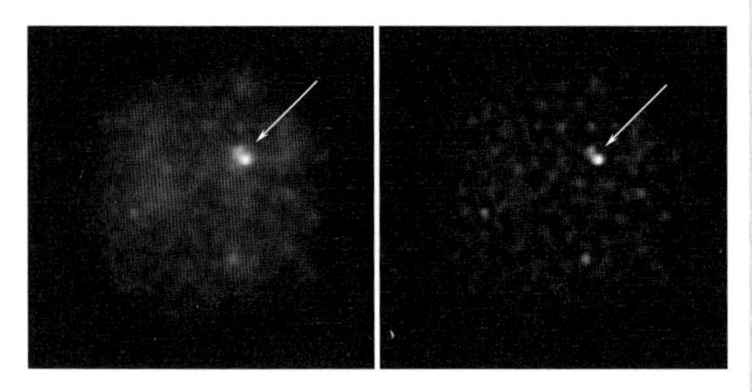

Left: the labeled cosmid probes are too close to be resolved (white arrow).

Right: after deconvolution, the contrast is improved and the two signals are discriminated.

Figure 6.21 2-D digital deconvolution of the image of a lymphocyte nucleus.

An example of 2-D image restoration is shown for two different FISH images: (i) *in situ* hybridization of cosmid probes in an interphase nucleus, and (ii) DAPI-stained metaphase chromosomes (in mouse). We can see the real improvement in resolution. The fluorescent signals can be resolved, while the DAPI banding of chromosomes is dramatically emphasized.

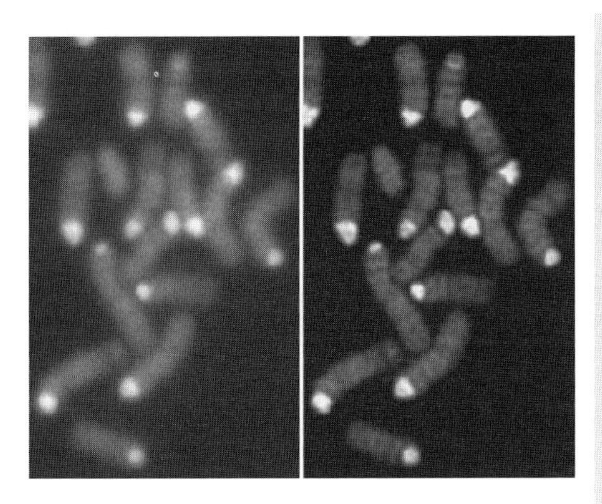

Left: raw image; chromosome G banding is hardly visible

Right: after deconvolution; the G banding is perfectly visible and allows karyotyping

Figure 6.22 2-D digital deconvolution of the image of mouse chromosomes labeled with DAPI.

6.2.4.2 3-D deconvolution

In confocal microscopy, deconvolution is not actually used to improve lateral resolution but rather to reduce the difference between lateral and axial resolution. In order to "deconvolve" confocal 3-D sets, it is necessary to measure the PSF of the microscope. This can be done by recording a series of confocal sections through an object whose size is perfectly known and which is smaller than the resolution power of the objective lens. Perfect objects for this purpose are plastic beads with a diameter of 0.15 μm and coated with a fluorochrome. Solutions of fluorescent beads with various diameters and type of fluorochromes are distributed by the main fluorescent probe manufacturers. In order to measure the PSF, we select an isolated bead in the preparation and collect a series of confocal sections through this bead. We filter out noise from the 3-D data set. This is repeated for nine other isolated beads. The obtained 10 3-D volumes are brought to registration by centering their data around the voxel of maximum intensity. Then, an average 3-D volume is obtained by merging the 10 different volumes after registration.

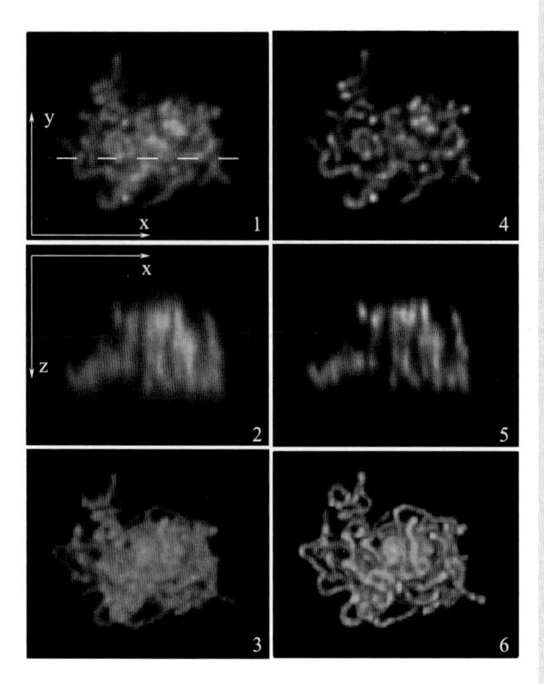

1 = Raw *xy* section through the chromosomes; the white dashes show the level of the *xz* section (2 and 4).
2 = Raw *xz* section; this image appears as blurred along the *z* axis; this is due to the poor axial resolution.
3 = 3-D reconstruction of the whole volume.
4 = Same *xy* section as 1 after 3-D deconvolution.
5 = Same *xz* section as 2 after 3-D deconvolution; the gain in axial resolution is obvious, though an axial smear is still perceptible.
6 = 3-D reconstruction of the whole volume after 3-D deconvolution.

Figure 6.23 Digital deconvolution of confocal series; DAPI staining of condensing chromosomes of *Xenopus laevis*.

As shown in Figure 6.23, 3-D deconvolution greatly improves axial resolution. In the raw *xz* image (Figure 6.23-2), all the structures seem to smear along the vertical direction, which corresponds to the *z* axis. However, although the axial resolution is improved by deconvolution, it is not possible to obtain a perfectly isotropic resolution. A seen in Figure 6.23-5, the axial resolution is poorer than the lateral one. All the profiles are still slightly smeared in the *z* direction.

6.2.5 Correction of Geometrical Aberrations

The last point to be dealt with concerns the possible distortion of the volume along the optical axis. During the acquisition of serial sections, the *Z* motion of the microscope stage can differ from the actual optical motion of the focal plane. Indeed, the objective lenses are designed to work at a given work distance and for a specific combination of refractive indices. That is, the objective is designed to work in water and/or oil, with a glass coverslip of a given thickness, and the refractive indices of the montage medium and of the sample should match one another and match the refractive index of the immersion liquid of the objective. If such conditions are not satisfied, the stack of sections is either shrunk or elongated in the *Z* direction.

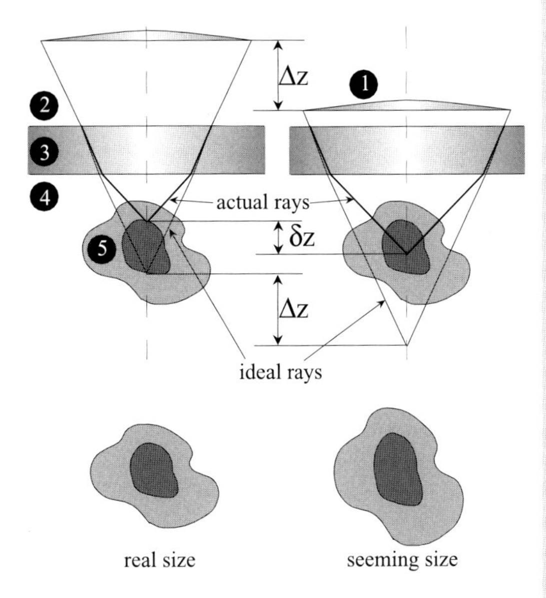

Top: optical path of ideal and actual rays in case of mismatch of the refractive indices

1 = Objective lens
2 = External medium (refractive index n_1)
3 = Glass coverslip (refractive index n_2)
4 = Embedding medium (refractive index n_3)
5 = Object (refractive index n_4)

When the objective lens is physically moved Δz, the actual optical motion (δz) is significantly smaller.

Bottom: the reconstructed object is distorted (right). It looks more elongated along the *z* axis than it actually is (left).

Figure 6.24 Geometrical distortions.

Although it is preferable to use a correct optical setting, it happens that the refractive indices cannot match or the thickness of the coverslip is not optimal (only few objective lenses have a ring for thickness correction). In such a case, it is mandatory to correct the geometrical distortion. This is simply obtained either "in-line" by correcting the *Z* sampling step of the microscope stage or "off-line" by digital resampling.

➩ Most of the available confocal systems offer an option to correct the *Z* step.

Chapter 7

Fluorescence Image Analysis

Yves Usson

CONTENTS

7.1 INTRODUCTION

The analysis of microscopy images is based on the measurement, in all points of an image, of the amount of either absorbed light (bright-field microscopy) or emitted light (fluorescence microscopy). Emitted light can be produced by fluorescent molecules that are specifically bound to some cellular components. Because all the cellular and background elementary points are systematically scanned, an image of a microscopic field can be converted to a numerical matrix that can be easily handled by a computer. The different steps in processing such a numerical matrix are:

⇨ Light absorption may be due to staining (absorbing dyes) or be created by interference between light rays (phase or differential contrast microscopy).

- Image acquisition by means of captors suited for large ranges of light dynamic as well as adapted to requirements of fine spatial resolution.
- Preprocessing to enhance the contrast of the image and correct for heterogeneous background illumination.
- Segmentation, which defines among the elementary points of the digital image, those which belong either to a cell or to any other subcellular component.
- Feature extraction that makes it possible to calculate various parameters, which describe the size, shape, photometry, color, and texture of the region or object of interest.
- Data analysis that makes it possible to find among the numerous calculated variables those which exhibit the significant variations in the context of an experiment or a pathology.

⇨ The image captor can be either analog or digital. In the first case, the analog data must be converted to numerical data by a frame grabber or a digital signal processor (*DSP*).

⇨ In fluorescence, a noise filtering step is often required and background correction is mandatory.

⇨ This is the most important stage in image analysis, since the quality of the measurements is directly determined by the quality of the segmentation masks.

⇨ A lot of dedicated statistical packages are available, which provides us with the basic analysis tools and some more sophisticated multivariate methods.

7.2 IMAGE ACQUISITION AND DIGITIZATION

7.2.1 Image Acquisition, Pixels

Digitization is the stage where the electrical signal of the image captor is converted to sequences of discrete numerical values. When these values are stored in the memory of a computer, they can be used for numerical calculations.

⇨ "Analog-to-digital" conversion is achieved by either a frame grabbers or a DSP. Nowadays, DSPs tend to supersede frame grabbers in most of the commercial image analysis systems.

0-8493-0817-8/01/$0.00+$1.50

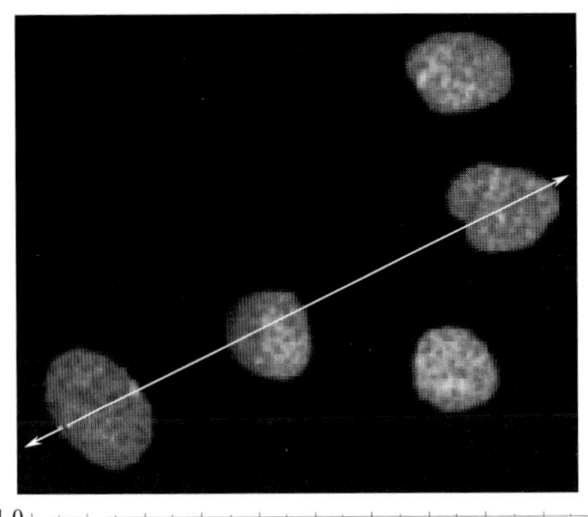

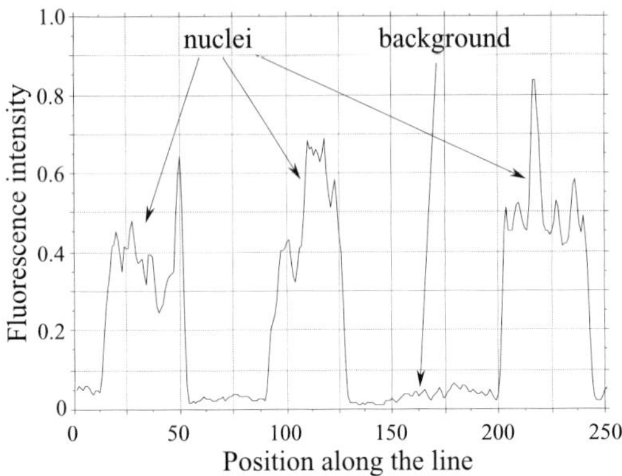

Top: fluorescence image of nuclei of human fibroblasts (DNA staining by TOPRO-3)

Bottom: fluorescence intensity profile measured in the image along the white line. As can be seen, the line traverses three nuclei. This results in three main plateaus in the intensity profile. The variations of intensity within the plateaus correspond to variations of chromatin condensation. The regions of low intensity between the plateaus correspond to the background. The slight intensity variations in these regions are due to low intensity auto-fluorescence, nonspecific labeling, and photon noise.

Figure 7.1 Nature of the image.

The initial image of the field of the microscope, the nature of which is continuous (Figure 7.1), is now a numerical array of n rows and m columns. At each crossing of a row and a column is stored a numerical value which expresses the amount of light collected at the same coordinates in the microscope field. Thus, one may consider that our original image has been divided into tiny area elements called *pixels* (compression for "picture element"), the average light intensities of which are coded by a discrete value (often called *gray level*) (Figure 7.2).

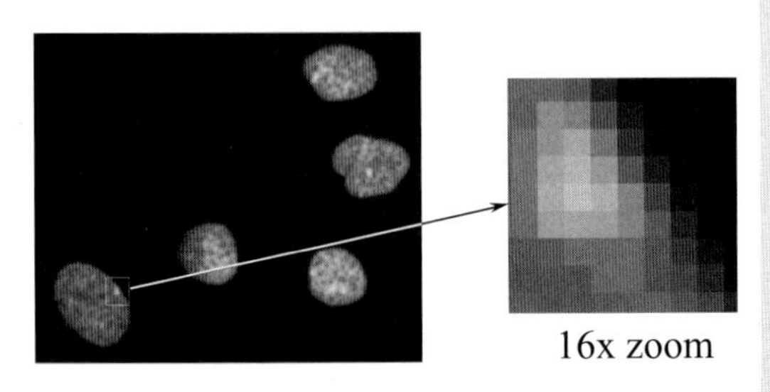

Left: digital image of nuclei of human fibroblasts

Right: ×16 enlargement of the region delineated by the white frame; the elementary image points or pixels are visible

Figure 7.2 Digitization.

7.2.2 Spatial Sampling

7.2.2.1 Discretization

The spatial sampling corresponds to the division of an image in elementary area units called pixels. In order to obtain a numerical image that preserves the smallest details of the microscopy image, the sampling must be performed with a high spatial frequency. In other words, the pixels must be as small as possible. For example, to take advantage of the resolution of the microscope, one must use a pixel size at least two times smaller than the actual resolution power of the objective lens (Nyquist's criterion, Shannon's theorem).

⇨ Discretization is the process by which a continuous space is converted to a discrete one.

⇨ The highest spatial frequency of a digital image depends on the pixel density. In the case of microscopy it is expressed in μm^{-1}. For example, if an image is composed of 1000 pixels and corresponds to a field of 100 µm, then the highest spatial frequency is 5 μm^{-1} (1000 pixels/(2 × 100 µm)).

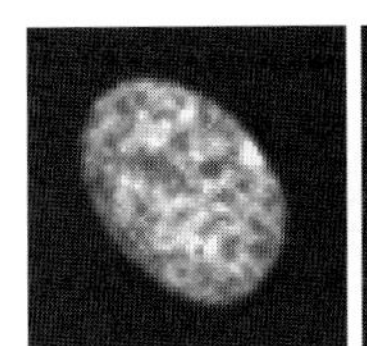
1 pixel = 0.2 µm

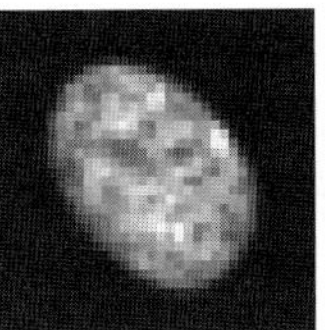
1 pixel = 0.4 µm

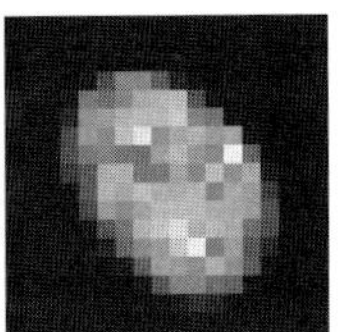
1 pixel = 0.8 µm

Left: digital image sampled with a high frequency (1 pixel = 0.2 µm)

Right: the same image sampled with a low frequency (1 pixel = 0.8 µm)

Figure 7.3 Spatial sampling.

Generally, the spatial sampling will be trimmed to the desired size of detail. As seen in Figure 7.3, when using a low spatial frequency (pixel = 0.8 µm), the finest features (condensed chromatin granules, Barr's body, etc.) are lost and a higher sampling frequency should be used (pixel = 0.2 µm). Furthermore, the shape of the nucleus becomes jagged and jeopardizes morphometrical analysis.

7.2.3 Intensity Sampling

7.2.3.1 Sampling

The numerical scale defines how the whole intensity dynamics of a signal will be converted in discrete numerical values (gray levels). It sets the range of detectable intensity values but also the intensity quantum between successive numerical values.

⇨ The dynamics of the signal corresponds to the greatest range of possible intensity values, that is the difference between the highest intensity and the lowest.

In practice, various digitization scales may be used. More frequently, a gray shade image is digitized with a scale of 256 gray levels linearly spaced from black to white. However, scales with more gray levels (1024, 2048, and 4096) are also used in cases where a greater dynamic precision is required (improvement of the signal-to-noise ratio). In some particular applications, the gray levels may not be linearly spaced (logarithmic scale, exponential scale, etc.). The variations of intensity and texture are correctly rendered with a 256 gray level scale, and they are poorly rendered with 8 gray levels.

⇒ 256 gray levels are sufficient to correctly analyze images of transmitted light microscopy.

⇒ 1024 and more gray levels are required to correctly analyze images of fluorescence microscopy in which very faint but significant fluorescence intensities can lie side by side with regions with very bright fluorescence intensities.

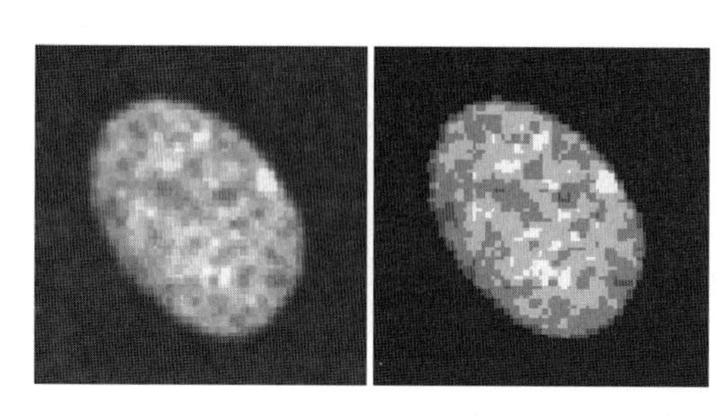

Left: image digitized with 256 gray level scale

Right: same image digitized with 8 gray levels

Figure 7.4 Numerical scale.

7.2.3.2 Gray level distribution

The examination of the gray level dynamics provides valuable information regarding the image content. The gray level distribution histogram offers a statistic for all the pixels of the image. The pixels are grouped in homogeneous classes of gray levels. In the gray level histogram of the fibroblast nuclei (*see* Figure 7.5), two populations of gray levels can be distinguished. These populations, which are roughly "Gaussian like," may be directly associated with particular zones or phases in the image. For example, the major population (gray levels between 0 and 50) corresponds to the darkest values in the image — in other words, those found mainly in the background of the image. The second population (levels 50 to 200) corresponds to the nuclei of the fibroblasts.

⇒ When using a linear scale for displaying pixel frequencies, it is very often difficult to distinguish the important gray level information. In fluorescence images, the relevant information (nuclear fluorescence, FISH signals) only represents a small area of the image (in general no more than 10%). Therefore, when building the gray level histogram, the background information becomes predominant and tends to hide the relevant fluorescence signal.

⇒ Using a logarithmic scale for displaying pixel frequencies helps to reveal the seldom represented but relevant gray levels (e.g., the nuclear pixels).

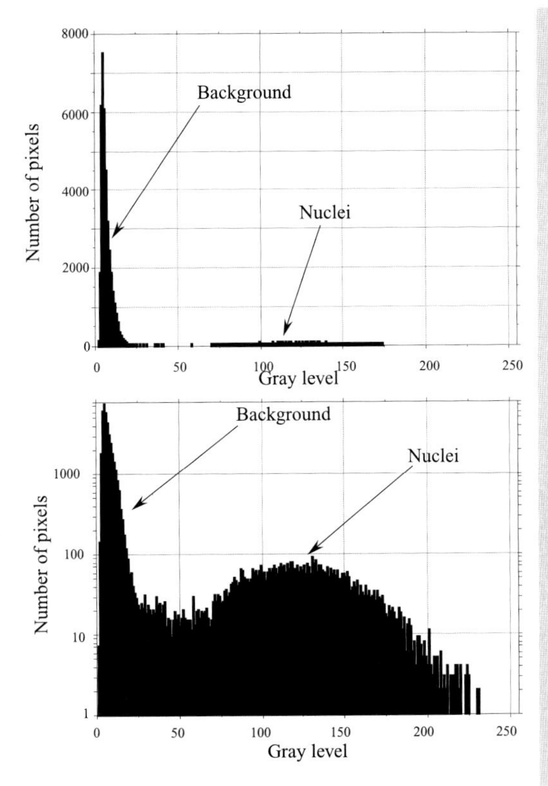

Top: gray level distribution of the image shown in Figure 7.1; the abscissa corresponds to the gray level from 0 (black) to 255 (white); the ordinate axis corresponds to the number of pixels with the same gray level; a main population of pixels (gray level 0 to 50) corresponds to the background pixels, while a smaller population of gray level (between 50 and 200) corresponds to the pixels of the nuclei

Bottom: use of a logarithmic scale for the ordinate axis improves the perception of the population of pixels that belongs to the nuclei

Figure 7.5 Gray level histograms.

7.3 IMAGE SEGMENTATION

7.3.1 Definition

The aim of image segmentation is to define and label the structures or regions of interest that must be analyzed. In general, segmentation algorithms are classed in two main categories often referred to as: the regional approach and the boundary approach. These methods generate binary images out of the gray level images. In these binary images, only two complementary classes of pixels are represented: the "object" pixels and the "background" pixels. The object pixels correspond to the regions of interest, and the background pixels correspond to all that is not a region of interest. The object pixels define binary masks that will be used for morphological and photometric measurements.

7.3.2 Binarization

7.3.2.1 Regional approach

As we saw previously, the analysis of the gray level histograms provides useful information on how a gray level image is organized. There is a link between the populations revealed by the gray level histogram (see Figure 7.5) and the different areas in the image. The dark gray level population is principally made of pixels that belong to the background of the image, while the bright gray level population corresponds to the fluorescent labeling of the nuclei.

⇨ Regional segmentation implies that the gray level image contains different areas in which the pixels share some common property. For example, the gray level of all pixels within an area oscillates around a typical value.

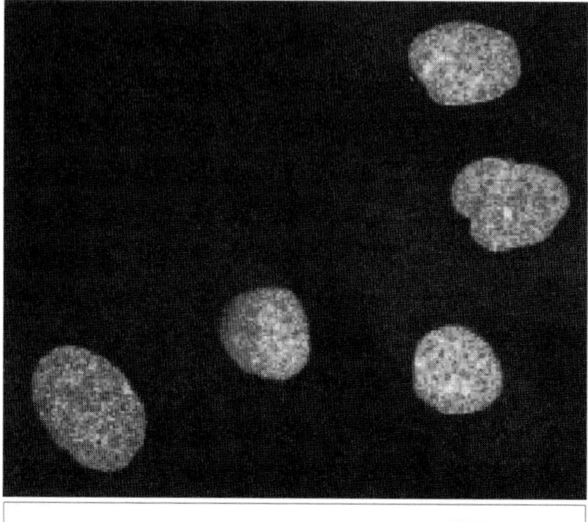

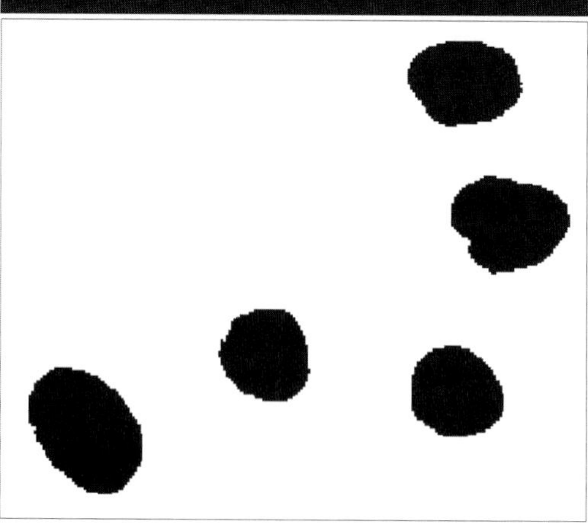

Top: the gray level image

Bottom: the binary mask, the "object" pixels are displayed in black while the background pixels are in white.

This segmentation result is obtained using a threshold value of 50 selected after analysis of the gray level histogram shown in Figure 7.5.

Figure 7.6 Segmentation, regional approach. Gray level thresholding.

The most trivial method for regional segmentation is "gray level thresholding." First, it consists in calculating the gray level histogram and identifying the significant gray level populations. Then, it is possible to select threshold values between the populations. For example (*see* Figure 7.5), a single threshold value (gray level 50)

⇨ Although trivial, it reveals itself as very efficient in a majority of image analysis applications and, in particular, in the case of fluorescence images.

can be selected between the population of nuclear pixels and the population of background pixels. The segmentation rule is very simple and can be stated as follows:

- All those pixels of gray level value greater than the selected threshold are considered "object" pixels.
- All those pixels with a gray level value smaller than the selected threshold are considered "background" pixels.

⇨ In our example, the nuclear pixels have a gray level value greater than 50 (selected threshold).

In practice, the selection of the segmentation thresholds can be done in an interactive way. The human operator may define the thresholds through a graphical user interface (GUI). He "plays" with a virtual slider and immediately sees the result on the computer screen. Although this procedure is simple and quick, it suffers from all the inconvenience of human decision making. That is, the result is very subjective and depends on the human operator. It will vary significantly depending on:

- The user and his/her ability
- The rank of analysis in a series of images; that is, an image analyzed at the beginning, in the middle, or at the end of a series
- The hour of the day and the degree of weariness of the user, and so on.

⇨ As a matter of fact, segmentation is a very sensitive stage in the image analysis process. As was stated before, all the subsequent measurements are conditioned by the quality of the segmentation. A slight variation of size or shape of the segmentation mask can introduce significant bias in the measurements. Therefore, it is essential to thoroughly assess and put to the test all the segmentation procedures.

In order to avoid these problems, the selection of the threshold can be done in a less biased manner by a computer. The "decision-making" process is handled by an algorithm which performs a statistical analysis of the gray level histogram to find the significant points, such as population modes and unbiased discriminant frontiers between populations. The main advantage of such an unsupervised algorithm is its great consistency: the segmentation of a given image will always give the same result.

⇨ Unbiased frontiers in the sense of Fisher's discriminant analysis; that is, the population boundaries are set in such a way that the populations are intrinsically homogeneous and as different as possible from each other.

7.3.2.2 Boundary approach

In some cases, the contrast between the structures to be detected can be low, and the regional approach may fail. However, it is still possible to distinguish frontiers, which will be detected using the boundaries segmentation approach.

⇨ In such a case, the objects of interest are not characterized by their inside content but rather by their edges.

Conversely to the regional approach, which relies on the spatial homogeneity of light intensity within areas, the boundary approach relies on the fast variations of light intensity between these areas. If we refer to Figure 7.1 or 7.7 (top) and concentrate on the intensity profile, it is noticeable that while the areas of the nuclei are mainly characterized by their average intensity, these are delimited by two abrupt variations of the intensity. Reading the profile from left to right, we see that on the left side of a nucleus there is a rapid rise of intensity from outside to inside the nucleus and on the right side, a fall of intensity from inside to outside.

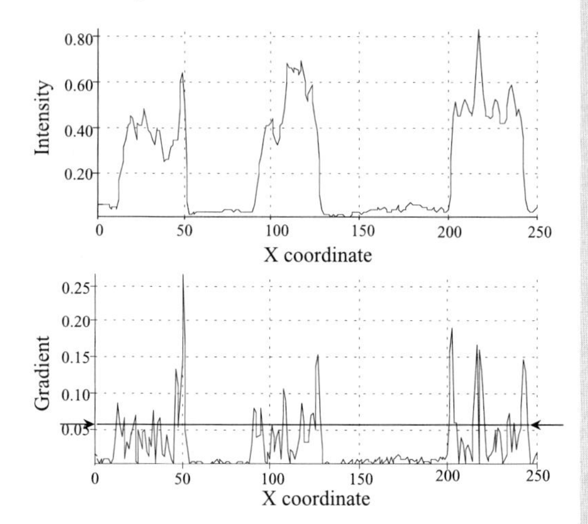

Top: fluorescence intensity profile through fibroblast nuclei (Figure 7.1)

Bottom: corresponding gradient intensity profile. The gradient at the x coordinate is calculated as the absolute difference between the fluorescence intensities at the $x - 1$ coordinate and at the x coordinate, respectively. The solid line (see arrows) corresponds to the gradient threshold above which pixels are considered boundary pixels.

Figure 7.7 Segmentation, boundary approach. Principle of edge detection by calculating an intensity gradient.

The detection of these rapid variations requires a transformation of the image. We must calculate a new image in which the values stored in pixels will represent the local intensity gradients. Different ways of characterizing intensity gradients can be used.

They rely either on the calculation of the first or the second derivative of the image. An example of such a transform is illustrated in Figure 7.7. In this figure, we calculate the absolute value of the first derivative. Therefore, the abrupt intensity variations at the edge of the nuclei (Figure 7.7, top profile) are converted into gradient picks (Figure 7.7, bottom profile). Then the edges of the objects can be detected by applying a gradient threshold: all those pixels whose gradient value is greater than the threshold will be considered "edge pixels" and the others will be considered "background pixels."

⇨ Calculating the derivative of an image may appear a complicated task to the non-mathematician. However, it is very simple since it consists basically in calculating the difference of intensity between a pixel and its immediate neighbor pixel. For example, the gradient values in the profile of Figure 7.7 are obtained with this simple formula:

$$G(x) = |I(x) - I(x - 1)|$$

In most of the available software, a collection of linear convolution filters (Sobel, Prewit, Laplacian, and others) is made available for this purpose.

These filters are all based on the calculation of the first or second derivative of the image but differ in the way they cope with photon noise. Figure 7.8 illustrates boundary segmentation using a Sobel's filter. The regions of the image where the gray levels vary slowly correspond to low gradients and appear dark. The regions where the gray levels vary rapidly, in other words, edges, appear bright. The brightness is proportional to the difference of potential of the gradient. Then, it is simple to apply a gray level threshold to the gradient image to obtain the binary mask of the boundaries.

➾ Since these filters aim at emphasizing the rapid variations of intensity, they are very sensitive to noise. As a matter of fact, noise is random by nature and is characterized by rapid variations from one pixel to another. Therefore, a straightforward gradient calculation would strongly emphasize noise. To avoid this problem, these filters include a low-pass filtering step in order to reduce noise.

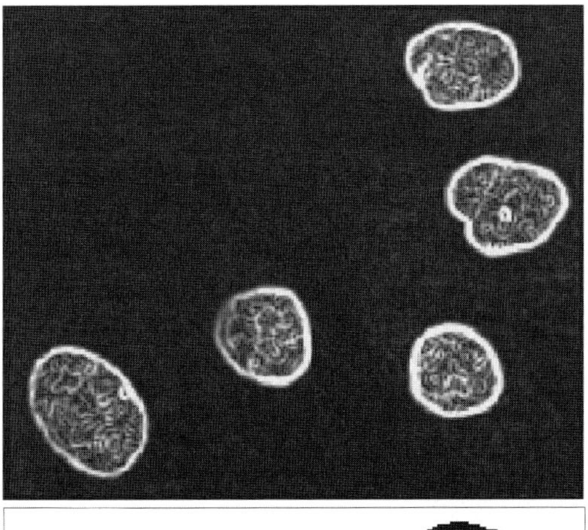

Top: detection of the boundaries with a gradient filter (basic Sobel's filter). The high gradients correspond to the bright pixels. The brighter values correspond to the nuclear edges. Other structures corresponding to the nuclear texture are emphasized by the gradient filtering.

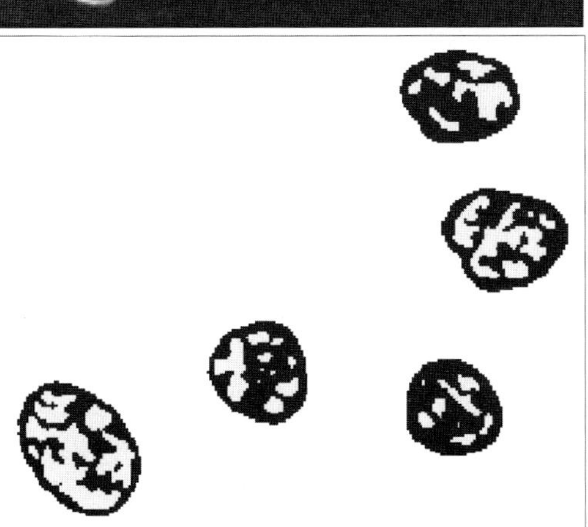

Bottom: thresholding of the gradient image. In this binary image, the black pixels correspond to the "edge pixels" and the white pixels correspond to the background pixels. The edge pixels inside the nuclei correspond to the frontiers between areas of heterochromatin and euchromatin.

Figure 7.8 Segmentation, boundary approach.

7.3.3 Shape Filtering

7.3.3.1 Basic operators

The binary images that are obtained with the above-mentioned methods are seldom satisfactory. They can be contaminated by numerous defects, such as small artifacts, holes in structures that are not supposed to be hollow, and they can have noisy contours.

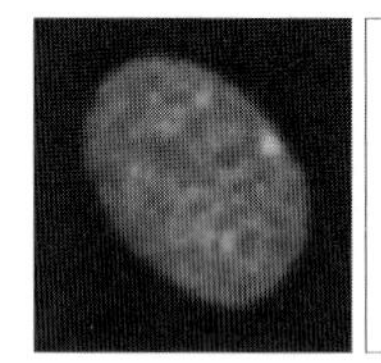

Figure 7.9 Effect of selecting a different gray level threshold. If the threshold value is set too high, holes are created within the mask of the nucleus. If the threshold is set too low, small structures (background noise) appear around the nucleus.

For example, in Figure 7.9, the binary mask of the fibroblast nucleus looks "moth-eaten" if the gray level threshold is set too high. The nuclear mask is hollow in the regions where chromatin is less condensed. Conversely, if the threshold is set to a lower value, the holes within the nucleus are filled up, but some "not relevant" structures appear outside the nucleus. Therefore, these segmentation masks cannot be used as such for further measurements. They need to be trimmed in order to remove the small artifacts, fill in holes in the masks, and eventually smooth out the contours of the masks. These various corrections are possible with the tools of mathematical morphology. The two basic operators of mathematical morphology are *erosion* (also called shrinking) and *dilation* (also called expanding). These operators can be combined to build more complex operators.

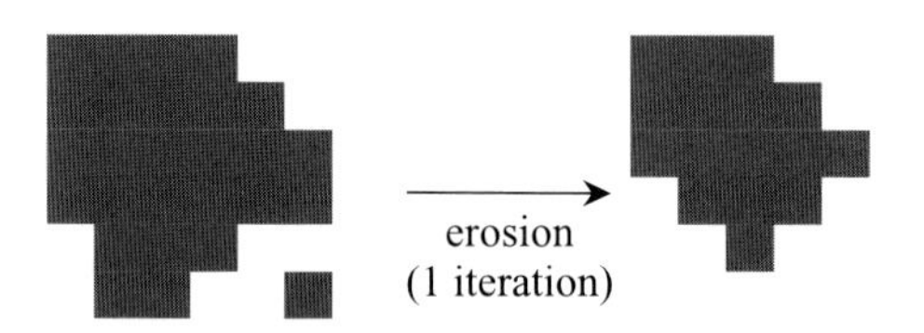

Figure 7.10 Morphological erosion. For each "object pixel," we look at its four direct neighbors. If at least one of these is a "background," the central "object" pixel is converted to "background."

Erosion is an operation that tends to shrink objects. To perform erosion, we consider that the object pixels have a value of 1 and background pixels have a value of 0. For every individual pixel (Figure 7.10), we record the value of the pixel and the values of its four direct neighbors (those sharing a face with the current pixel). Among these values we select the smallest and assign this value to the current pixel in the destination image. In a binary image, this process is similar to converting to background pixels those object pixels that are directly in contact with the background. In other words, the object pixels belonging to the border of the object are transformed to background pixels.

➯ For example, let us consider a pixel located in the center of a big object (Figure 7.10). It is surrounded by pixels with a value of 1. Thus, the smallest value is 1 and the current pixel is left unchanged. Now, let us consider the isolated object pixel or any of the border pixels of the big object. The minimum value found in the neighborhood is 0. Therefore, the current pixel will eventually be set to 0.

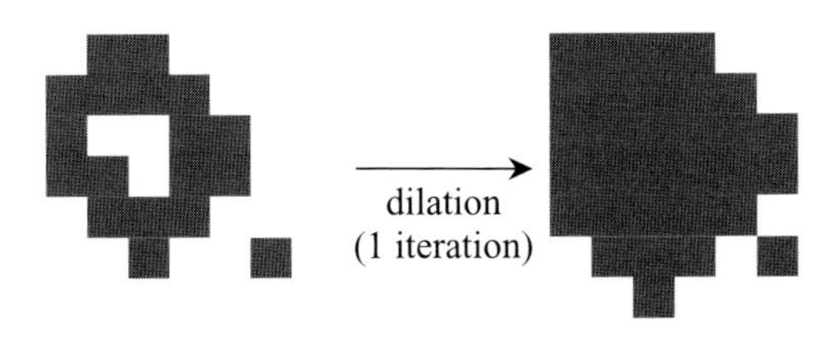

Figure 7.11 Morphological dilation. For each "background" pixel, we look at its four direct neighbors. If at least one of these is an "object" pixel, the central "background" pixel is converted to "object."

Dilation is the operation complementary to erosion. For every individual pixel (Figure 7.11), we record the value of the pixel and the values of its four direct neighbors (those sharing a face with the current pixel). Among these values we select the greatest and assign this value to the current pixel in the destination image. In a binary image, this process is similar to converting to object pixels those background pixels that are directly in contact with the object. In other words, the background pixels belonging to the border of the background are transformed to object pixels.

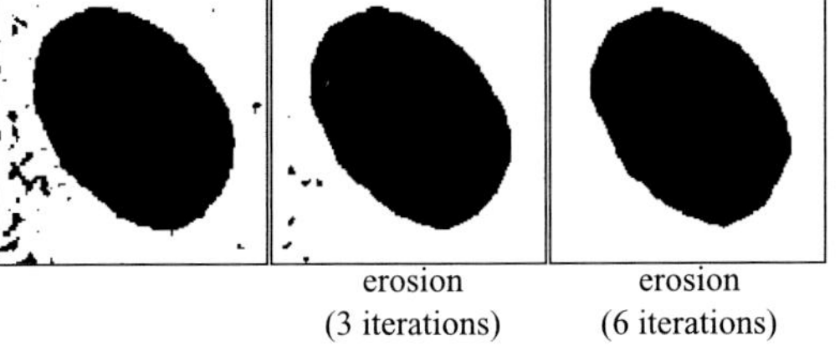
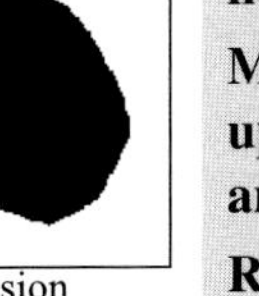

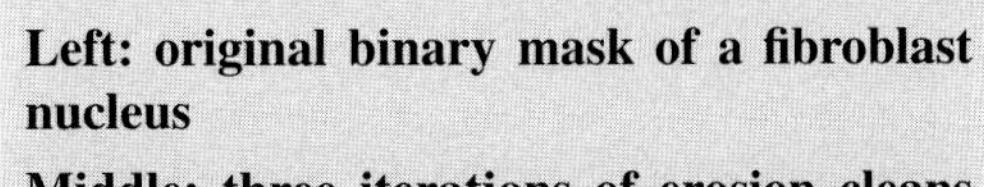

Left: original binary mask of a fibroblast nucleus

Middle: three iterations of erosion cleans up the mask but small external structures are not discarded

Right: final mask after five iterations

Figure 7.12 Iterative erosion.

The erosion and dilation operators may be applied repeatedly. For example, an erosion may be repeated three consecutive times in order to erase small structures whose radius (or half-width) is smaller than or equal to three pixels. The number of iterations may also be called the depth of erosion. In the same manner we may repeat dilation three times in order to fill in holes whose radius (or half-width) is smaller than or equal to three pixels. The number of iterations may also be called the depth of dilation.

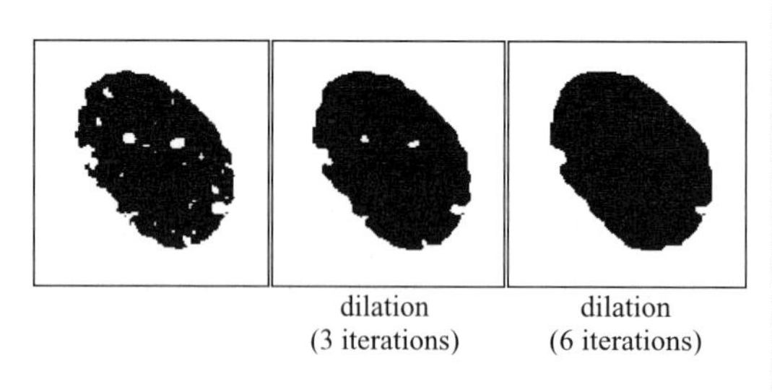

Left: original binary mask of a fibroblast nucleus

Middle: three iterations of erosion cleans up the mask but some holes still remain

Right: final mask after five iterations

Figure 7.13 Iterative dilation.

Erosion and dilation may also be successively applied to achieve more complex shape filtering such as morphological closure and morphological opening.

7.3.3.2 Combined operators

The closure operation consists in applying dilation with a specified depth and then applying erosion to the same depth. The dilation step makes it possible to fill in holes with a given radius. For example, in Figure 7.14 all the holes inside the nuclear mask have been filled up after six iterations. However, the size and shape of the nuclei are overestimated. The following erosion with the same depth restores the original dimensions of the nucleus, but the holes are not regenerated. It must be noted too that the contour of the nucleus is smoother; the small irregularities have been pruned out by the closure.

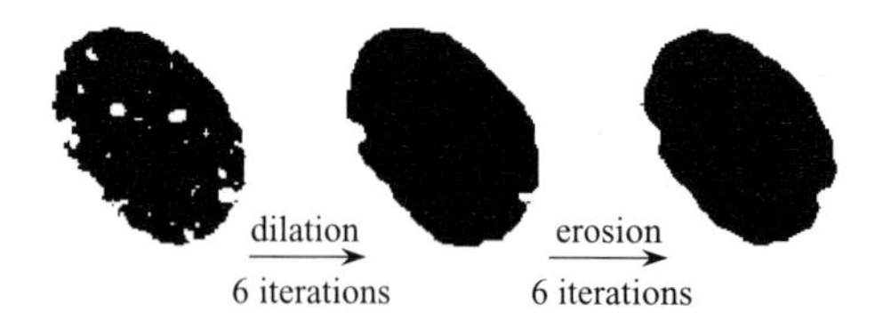

Figure 7.14 Combined morphological operators: Closure. Closure is obtained by applying dilation then erosion. The depths of dilation and erosion must be the same.

The opening operation consists in applying erosion with a specified depth and then applying dilation to the same depth. The erosion step makes it possible to eliminate isolated small structures (coloration artifacts). For example, in Figure 7.15 the small artifacts around the nucleus have disappeared after six iterations. However, the nuclear mask looks "skinny" and its size is underestimated. Dilation to the same depth restores the original dimensions of the nucleus. It must be noted too that the contour of the nucleus is slightly different from what it was before the opening.

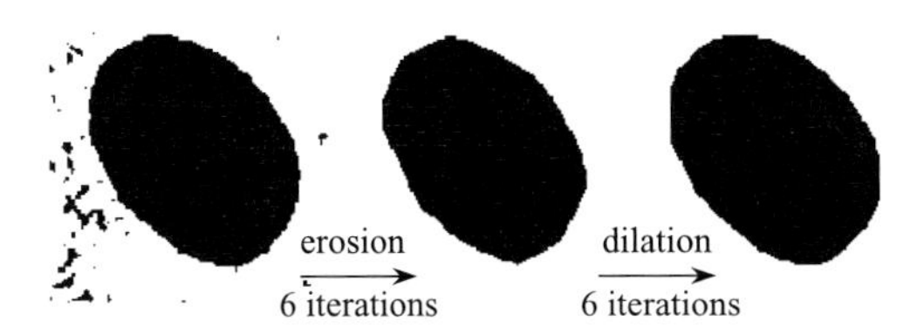

Figure 7.15 Combined morphological operators: Opening. Opening is obtained by applying erosion then dilation. The depths of erosion and dilation must be the same.

7.3.4 Connected Component Labeling

7.3.4.1 Principle

A final step is required before feature extraction can be performed. The thresholding and the morphological filtering allowed us to define regions of interest. However, in most cases it is useful to distinguish different entities within the binary image in order to obtain separate measurements. For example (*see* Figure 7.6), if one wants to know the average area of a fibroblast nucleus, it is necessary to extract, one by one, the masks of the nuclei and perform the measurement. This is possible if the binary image has been labeled.

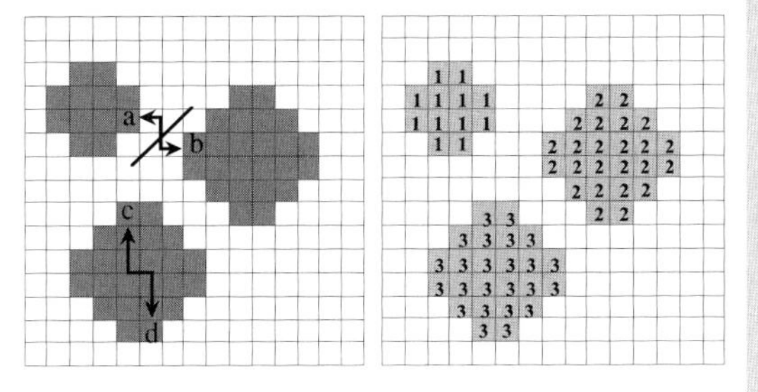

Left: binary mask to be labeled. Three different objects can be isolated. No continuous path can be found between pixels *a* and *b*. A continuous path is found between pixels *c* and *d*.

Right: image of the labels. A new image is produced by the labeling algorithm. All those pixels that belong to the same set are given the same number.

Figure 7.16 Set of connected components.

Connected component labeling consists in giving to each separated object region (or set of connected components) a label; that is, a unique number is assigned to all the object pixels of that region (Figure 7.16). The basic definition of a set of connected components is the following: let us consider two object pixels; these two pixels belong to the same set of connected components if a continuous path between them can be found that is only stepping on object pixels. If no path can be found without passing along background pixels, then they belong to different sets.

7.3.4.2 Connexity rules

The notion of connexity corresponds to the neighbor relationship between a pixel and the surrounding pixels. In general, two main connexity rules are used:

- "4 connexity," where the lateral pixels are considered candidate neighbors: those that share a common face with the central pixel (above, under, left, and right)
- "8 connexity," where the lateral and diagonal pixels are considered candidate neighbors: those that share either a common face or a corner with the central pixel

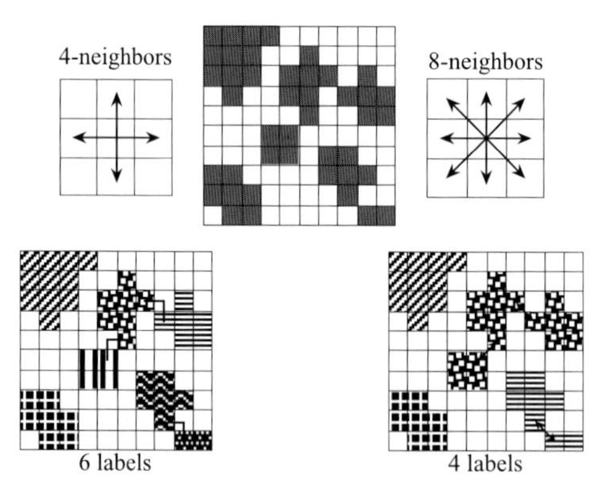

Figure 7.17 Connexity rules. Labeling the same binary image using different connexity rules gives rise to different results. For example, when using a "4 connexity" rule, we obtain six different objects. When using the "8 connexity" rules, we obtain only four different objects.

7.4 QUANTITATION

7.4.1 Morphometry

Morphometry consists in extracting features or measurements that are characterizing the size, shape, and structural organization of an object. After extraction of the mask of a labeled object, it is possible to measure a collection of features (Figure 7.19). For example, the area of an object is simply obtained by counting the number of pixels that belong to that mask. If the acquisition systems have been thoroughly calibrated (i.e., if the size of a pixel has been measured), it is easy to obtain a direct area measurement expressed in μm^2. One can also calculate the position of the center of mass of the object. It is particularly interesting in mapping applications when one needs to know the distribution and position of particular cells with reference to a particular histological structure (border of a tissue, lumen of a gland, etc.) or to study the relative spatial organization of cells in normal and pathological tissue.

⇨ If the microscope is equipped with a motorized stage, the value of the center of mass can be stored in the memory of a computer together with the current coordinates (x,y) of the stage. This makes it possible later on to retrieve cells and objects on the slide by recalling their locations from a database.

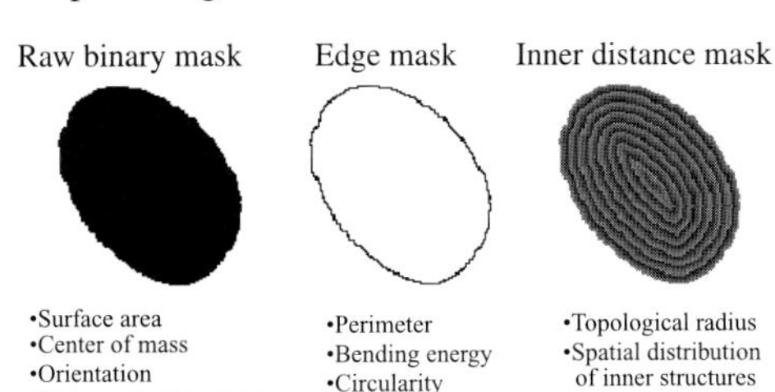

Left: features derived from the mask of the object
Middle: features extracted from the edge of the mask
Right: topological features derived from the map of inner distances from the edge

Figure 7.18 Morphometry.

Other types of measurements can be obtained after the extraction of the contour of the mask. For example, the perimeter is obtained from the sequence of pixels that belong to the contour. Two consecutive neighbors sharing a face (lateral neighbors) are separated by a unit distance (1), while two consecutive pixels sharing a corner (diagonal neighbors) are separated by a distance equal to the square root of 2 (1.414). Here again, if the lateral size of a pixel has been calibrated, the perimeter measurement may be expressed in µm. Other shape measurements, such as the bending energy of the contour, the roundness (or form factor), can be obtained.

⇨ The bending energy expresses how the contour of an object could be reconstructed by assembling arcs of various lengths and radii.

⇨ The roundness or circularity compares the shape of the object with an ideal circle. This is expressed by the formula $R = P^2/(4\pi A)$, where P is the perimeter and A is the area. The circle is a unique shape in the sense that it encloses the greatest possible area for a given perimeter. Thus, the roundness of a circle is 1. For any other shape, this value is greater than 1.

It is also possible to extract topological features that describe how things are organized in a closed space. For instance (see Chapter 8), it may be useful to quantify the spatial organization of chromatin within a nucleus because it is a signature of replication activities during interphase. Another interesting application is the study of the spatial distribution of some genes within the nucleus by FISH labeling. Most of the topological features rely on distance measurements with reference to some particular structure. In a nucleus, the nuclear membrane is a very interesting reference since it can be easily identified and it represents the natural boundary between DNA and the cytoplasm. All messengers and transcripts must pass through this surface. Given a binary mask of the nucleus, it is possible to calculate a topological map of the nucleus (Figure 7.18).

The area of the nucleus is divided in "onion skins." Each layer or "skin" corresponds to a unique inner distance within the nucleus. In other words, all the pixels within a layer are the same distance from the edge of the mask. Then it is easy to locate a FISH signal in the nucleus by looking at the intersection of the center of mass of a signal with an "onion skin."

➪ A distance map can easily be generated by applying iterative erosion. After each erosion step, the pixels that were removed from the mask are stored in the map with a label corresponding to the current number of iterations. The process is iterated until no pixel is left in the binary mask.

7.4.2 Photometry

The photometric information (gray levels) gathered in the images can be used to measure the amount of substances in structures. In fluorescence microscopy, and in nonsaturating conditions, the light intensity is expected to be proportional to the amount of fluorochrome bound to the labeled substance. This is true only if the fluorescent labeling relies on a stoichiometric hybridization or immunochemical process. In such a case, the amount of light collected in a pixel may be related to the amount of the labeled molecule. Photometry takes advantage of light intensity to provide quantitative measurements of molecule amounts within cells and nuclei.

In order to measure a significant light amount, it is important to pay strict attention to:

- The stoichiometry of the labeling process
- The evenness or homogeneity of illumination in the microscope field
- The amount of excitation light that must be set to avoid saturation of the fluorophores and photobleaching
- The screening time of the slide before acquisition; the shorter the better

➪ Here we must temper our enthusiasm, because the conversion from fluorescence intensities to molecule amounts is not that straightforward. A lot of extrinsic factors impair the efficiency of fluorescence emission (fading, quenching, energy transfer, spectral shift). Although at first glance it looks simple, the measurement of fluorescence actually demands great care, and the results must be interpreted with caution. For example, a variation of fluorescence smaller than 10% between two nuclei or cells is hardly interpretable.

➪ In fact, it would be better not to observe the slide prior to image acquisition!

Once these precautions have been taken, we can start to measure fluorescence intensities. Two processes are required to obtain an image suitable for quantitation:

- Dark current subtraction
- Field equalization or correction of field homogeneity

The dark current subtraction is a stage required for image captors such as cooled CCD cameras (*see* Chapter 5). The dark current is a nonrandom perturbation of the intensity signal that is due to the slight differences in the electrical characteristics of the photodetector elements of the CCD chip. These slight defects, which are not relevant in the case of transmitted light microscopy, become significant in fluorescence microscopy where the capture time is greatly increased in order to cope with random photon noise. As a matter of fact, these defects, which nature determines, tend to be emphasized by time integration.

➫ Common CCD chips are made of an array of 1024 × 1024 photodetectors. It is quite impossible to produce "0-defect" chips. This would suppose an error rate smaller than 0.001%!

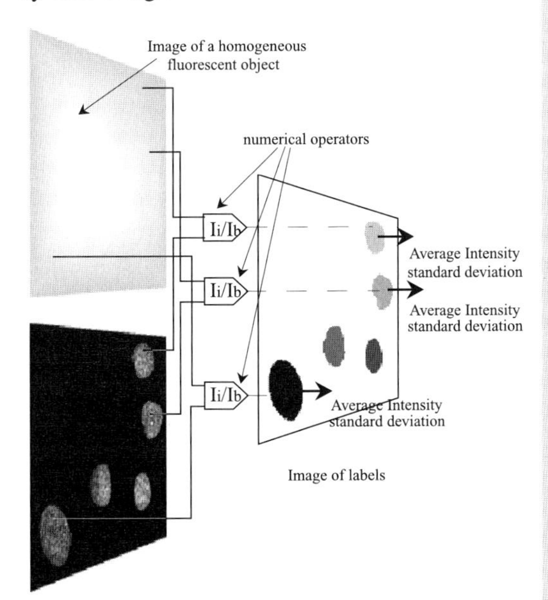

Figure 7.19 Photometric measurements. The fluorescence intensity is measured pixel by pixel. In the case of fluorescence images, the dark current of a CCD camera must be subtracted from the recorded image. Then the image must be corrected for illumination defects. For this purpose, a reference image is recorded, using a bright and constant fluorescence object that completely fills the microscope field. This image is the reference of homogeneity. Then, the image of the nuclei can be recorded, and the fluorescence value of each pixel is divided by the value of the corresponding pixel in the image of reference. Then the total and average fluorescence can be measured through the segmentation masks of the nuclei.

Dark current subtraction is obtained by recording the image provided by the CCD chip when the shutter of the camera is closed. In such a "dark current" image, no photon ever reached the chip; therefore the variations of intensity stored are due only to chip irregularities. The correction is straightforward. It consists in subtracting, pixel by pixel, the value of the dark current image from the value of the fluorescence image.

➫ The integration time must be identical to that used for normal exposure of the fluorescence image.

➫ Modern CCD cameras now integrate an "on chip" dark current subtraction in the acquisition process.

The next point we must address is field equalization or correction of field homogeneity.

In order to compare the fluorescence emitted from two different points of microscope field, it is essential that these receive exactly the same amount of excitation light. While such a condition is easily achieved in a CLSM where the laser beam scans the microscope field, it is quite impossible to obtain a perfectly even illumination in a conventional fluorescence microscope. Fortunately, it is possible to compensate for this. The idea is to measure the local illumination intensity on all points of the field and use this information to correct the measured fluorescence point by point. The reference image is obtained by recording the fluorescence of an object that completely fills the microscope field and which emits a perfectly homogeneous fluorescence over its whole surface. There are three kinds of fluorescence objects that can be used to generate this reference image:

➾ Each point in the microscope field is illuminated by the same laser point source and receives the same amount of excitation light.

➾ Each point is illuminated by a different part of the light source (xenon bulb) and receives a slightly different amount of excitation light.

- A solution of concentrated fluorophore spread between a coverslip and a glass slide
- Slide of uranyl–glass
- Pieces of Plexiglas dyed with fluorescent colors

➾ A concentrated solution of FITC mounted between glass slides is very efficient.

➾ The possession of such special glass may be difficult (or illegal) in some countries because of its slightly radioactive nature.

➾ These can be found in specialty shops selling plastic objects. A very money-saving alternative is to buy cheap red, green, or yellow fluorescent plastic rulers and cut test slides from them.

Chapter 8

Examples of Quantifications

Xavier Ronot
and Yves Usson

CONTENTS

8.1 CHROMATIN TEXTURE ANALYSIS IN LIVING CELLS

8.1.1 Principle

The nucleus of the living cell is a complex structure. Quantification of nuclear organization makes it possible to establish the state of DNA in the nucleus that provides an indirect representation of chromatin organization (condensation, topographical distribution).

Most studies have been made on fixed cells, i.e., on cell snapshots.

⇒ Nuclear properties are modified during the fixation steps, making them different from those in nuclei of living cells.

The existing methods used to quantify nuclear organization via texture analysis are based on the statistical or probabilistic assessment of the gray level distribution in the image.

⇒ These methods can discriminate between different populations but do not provide information about the visualized appearance of the initial image.

The application of a specific, stoichiometric, and nondisturbing method for DNA is described. This staining allows the real-time analysis of nuclear properties associated with the cell cycle phase of living cells and nuclear texture.

This approach has the advantage of providing parameters that can be directly interpreted by the user in terms of heterogeneity, granularity, condensation, and margination.

⇒ Since these parameters give an accurate representation of the chromatin distribution within the nucleus, their measure is particularly well suited to the continuous analysis of chromatin distribution at the population level as well as the individual level.

8.1.2 DNA Content and Morphological Measurements

On each nucleus, integrated fluorescence and average fluorescence intensities are measured. Morphometric measurements consist of area estimation.

Due to the stoichiometry of staining, integrated fluorescence intensity is directly related to the DNA content. Two quality control parameters were computed on collected data: the stoichiometric index (SI) and the G0/1 coefficient of variation of fluorescence intensities in the G0/1 phase.

0-8493-0817-8/01/$0.00+$1.50

SI can be derived from intensities according to the following formula:

$$SI = \frac{Fluo_{G2/M}}{Fluo_{G0/M}}$$

The coefficient of variation (CV) was computed on the G0/1 peak of the DNA content histogram. This parameter is related to the variability of measured intensities, either due to the staining procedure or to the imaging process.

⇨ The coefficient of variation is given by the ratio of the average gray level vs. its standard deviation.

8.1.3 Nuclear Texture Measurement

8.1.3.1 Principle

The method consists in segmenting the nucleus in a few regions sharing similar gray level values. Gray levels (256) are distributed in three different classes as compared to the mean gray level:

- The dark pixel class (*B*), which corresponds to pixels that display a gray level lower than the mean value interval measured on each object
- The gray pixel class (*G*), which corresponds to pixels that display gray levels included in the mean value interval
- The bright pixel class (*W*), which corresponds to pixels that display gray levels higher than the mean value interval

⇨ $Fluo_{G2/M}$ and $Fluo_{G0/1}$ are the mean values of fluorescence intensities for the G2 + M and G0/1 peaks displayed on the DNA content histogram.

⇨ An SI value of nearly 2.00 is associated with stoichiometric staining.

⇨ *See* Figure 8.1.

⇨ When measuring DNA content in living cells, samples with CV values lower than 15% are considered acceptable.

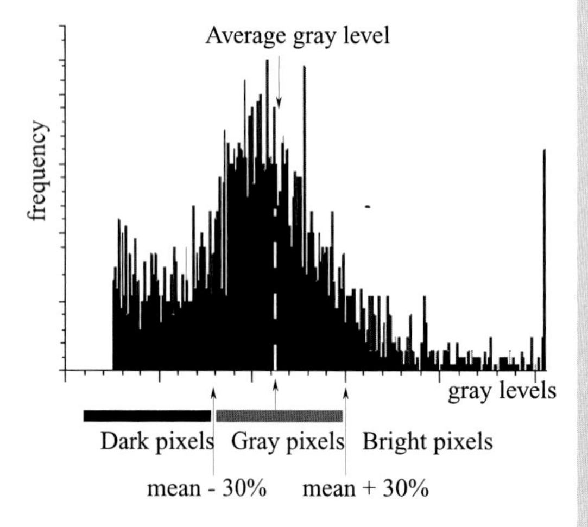

Figure 8.1 The mean value interval. MV = {mean – T, mean + T} where mean is the average gray value computed using all gray levels contained in the nucleus; T is an arbitrary threshold.

8.1.3.2 Chromatin condensation — heterogeneity and granularity parameters

In the three-level image, each gray level is considered a label. Measures are defined for nuclear heterogeneity and granularity, based on the relative number of pixels contained in each class. Heterogeneity is defined by the following formula:

$$Heterogeneity = \frac{N_W + N_B}{P}$$

⇨ N_B is the number of pixels in class B, N_W the number of pixels in class W, and P the total number of pixels in the nucleus.

⇨ A uniformly gray nucleus will yield a value for *Heterogeneity* of 0, while a completely heterogeneous nucleus, displaying all pixels either dark or bright, will yield a value of 1.

The granularity parameters, *clump* and *condensation*, are computed directly on the three-level image of nuclei. Such a labeled image is observed through a mesh. The mesh window is moved along all possible positions on the image, with no overlapping. At each position, the numbers of bright and dark labeled pixels (respectively, N_W and N_B) are scored and the absolute difference is computed. The *clump* parameter, which represents the size distribution of the "grains," is calculated as follows:

$$Clump = \frac{\sum_{mesh} |N_{Wmesh} - N_{Bmesh}|}{N_B + N_W}$$

The *clump* value increases as a function of the spot size, with a value close to 1 for images containing grains that are much larger than the mesh window size.

⇨ This parameter is sensitive to the absolute size of the spots in the nucleus. If these spots are small in size, the *clump* parameter value will be low, with a value close to 0 for images containing grains that are smaller than the mesh size.

Another parameter, *condensation*, is related to granularity. It takes into account the fraction of large spots as compared to the total nuclear area. This parameter is defined by the following formula:

$$Condensation = \frac{\sum_{mesh} |N_{Wmesh} - N_{Bmesh}|}{P}$$

8.1.3.3 Chromatin margination

The margination parameter is dependent on the tendency of chromatin to be located preferentially at the nucleus periphery. On the initial gray-level image, the average fluorescence intensity per pixel is measured successively in concentric rings that begin at the outer boundary and move inward to the center of the nucleus. The successive rings are obtained by iterative erosions of the binary mask of the nucleus. On the basis of average intensities, the second radial moment is calculated:

⇨ See Chapter 7, Section 7.3.3 and 7.3.4.

$$I_2 = \frac{\sum_{r=0}^{R-1} r^2 m(r)}{\sum_{r=0}^{R-1} m(r)}$$

⇨ $m(r)$ is the average intensity per pixel at a given radial position r, and R is the maximum radius. In the present case, radii are replaced by ring numbers. (R is the maximum ring number corresponding to the outer boundary of the nucleus.)

The nuclear margination is calculated as follows:

$$Margination = \frac{I_2}{R^2}$$

⇨ Values of *margination* are included between 0 and 1, with a value of 1/3 for an image with uniformly distributed staining.

8.1.4 DNA Content Measurement

On living cells, the use of Hoechst 33342 at the concentration of 10 μ*M* provides a stoichiometric staining. These staining conditions provide a right value of SI (1.98) and coefficient of variation (4.9%).

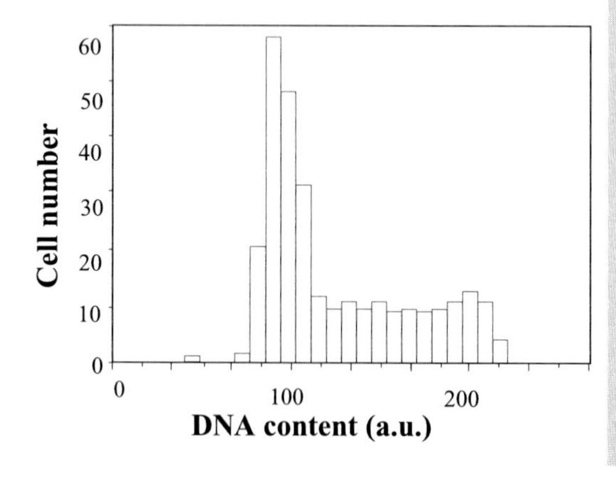

⇨ These results show that the measured fluorescence intensity is proportional to the DNA content of the analyzed nuclei.

Figure 8.2 DNA content distribution.

The cell cycle phase of individual nuclei could be determined before measuring texture characteristics on previously stored images.

8.1.5 Texture Measurements as a Function of Cell Cycle Phase

Texture measurements were performed on the whole cell population in order to assess the evolution of texture-related parameters during the cell cycle progression.

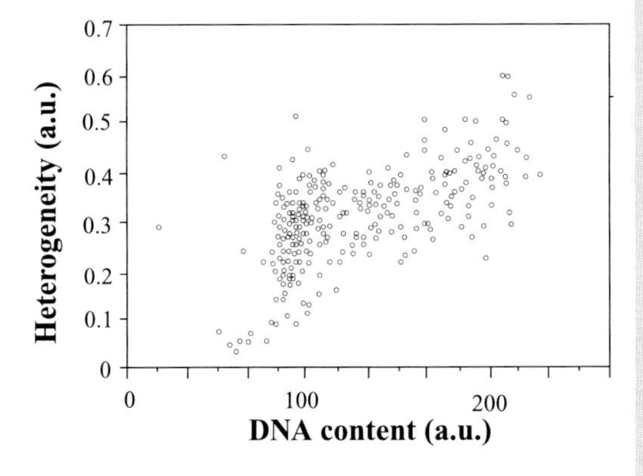

⇨ Heterogeneity increases with cell cycle progression.

Figure 8.3 Biparametric distribution of heterogeneity vs. DNA content.

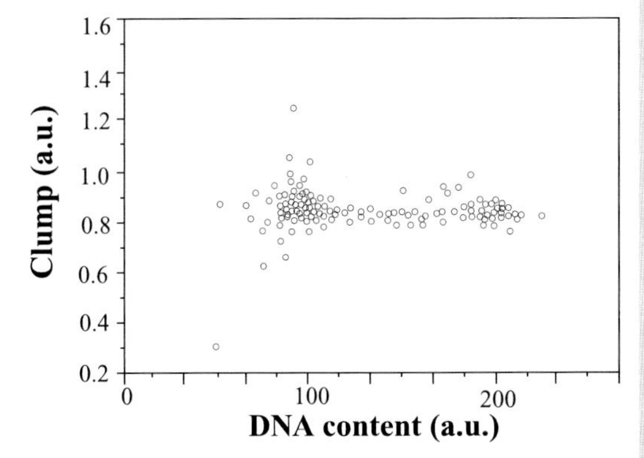

⇨ Clump parameter is not dependent on the cell cycle phase.

Figure 8.4 Biparametric distribution of *clump* vs. DNA content.

The organization of the chromatin varies with the progression of cell from G0/1 phases to the G2 + M phases of the cell cycle. *Heterogeneity* and *condensation* increase and *margination* slightly decreases, while *clump* remains constant.

The measure of texture parameters on nuclei of living cells can be successfully used to describe the evolution of chromatin organization during the cell cycle.

8.1.6 Material — Products / Solutions

8.1.6.1 Material

- Microscope
- Camera
- Image analysis software
- Glass chamber slides

⇒ The emitted fluorescence is supplied with excitation (365 nm), dichroic (395 nm), and long-pass (450 nm) optical filters.
⇒ SIT or CCD
⇒ For image acquisition and processing

8.1.6.2 Products/Solutions

- Culture medium
- Fetal bovine serum
- PBS 1x
- Hoechst 33342

⇒ Use PBS without Ca^{++} and Mg^{++}
⇒ Hoechst 33342; Stock solution: 1 m*M*/mL in H_2O, stored at 4°C

8.1.7 Protocol

8.1.7.1 Cell preparation

- Eliminate culture medium
- Wash the cells with PBS **5 min**
- Add serum free culture medium **30 min at 37°C**

⇒ Repeat three times

8.1.7.2 DNA staining

- Add Hoechst 33342 **30 min at 37°C**
- Eliminate staining solution
- Wash three times with PBS **5 min**
- Add complete culture medium **7 min**

⇒ Use serum-free culture medium containing 10 µ*M*/mL fluorochrome
⇒ Avoid light exposure
⇒ Avoid light exposure

8.1.7.3 Image acquisition

Fluorescence intensity is calculated on each pixel in arbitrary units according to the following formula:

$$Fi = \frac{V_m - V_e}{V_{ref}} \times 2.55$$

⇒ V_m: the measured value, V_e: value that corresponds to an empty field analyzed under same conditions, and V_{ref}: intensity of the pixel of interest measured on a homogeneous fluorescent reference.

- Select each microscopic field
- Acquire the images
- Preprocess them (empty field subtraction and image division by the homogeneity reference)
- Process to image segmentation by gray-level thresholding
- Store individual objects as separated image files until analysis

8.1.7.4 DNA content and morphological measurements

- On each nucleus measure:
- Integrated fluorescence
- Average fluorescence intensity
- Nuclear area

8.2 STUDY OF 3-D DISTRIBUTION OF AgNORs IN NUCLEI

8.2.1 Introduction

Silver staining of nucleolar acidic proteins is an interesting tool to discriminate malignant cells from benign cells in tumor pathology. Most of the existing methods to study the distributions of AgNOR proteins in nuclei are based on two-dimensional image analysis. However, confocal laser scanning microscopy brings a new insight by extending this study to three dimensions. In this chapter, we propose a quantitative method to discriminate cells belonging to three different leukemic cell lines by the characterization of the 3-D distribution of NORs in cell nuclei.

⇨ The silver precipitate of these nucleolar acidic proteins (NORs) is known under the name AgNOR.

Basically, the 3-D image analysis uses information from two different sources:

- The nucleus, which defines the measurement space
- The AgNORs

⇨ Fluorescence of propidium iodide is imaged in CLSM.
⇨ The silver aggregates are imaged in laser scanning transmitted light microscopy.

8.2.2 3-D Image Acquisition

8.2.2.1 Nucleus data set

The 3-D data set of the propidium iodide (PI) fluorescence (i.e., DNA labeling) is collected with an LSM410 confocal microscope (Carl Zeiss) fitted with a 63×, NA 1.4, Plan Apochromat oil-immersion objective lens. The PI fluorescence is excited with the 514 nm line of a helium–neon laser. A dichroic mirror with a 560 nm split wavelength is used to reject the excitation light and let the emitted fluorescence reach the detector. A 590-nm long-pass filter is placed in front of the detector in order to select PI fluorescence.

⇒ With this setting a lateral resolution of 0.22 µm and an axial resolution 0.4 µm are expected.

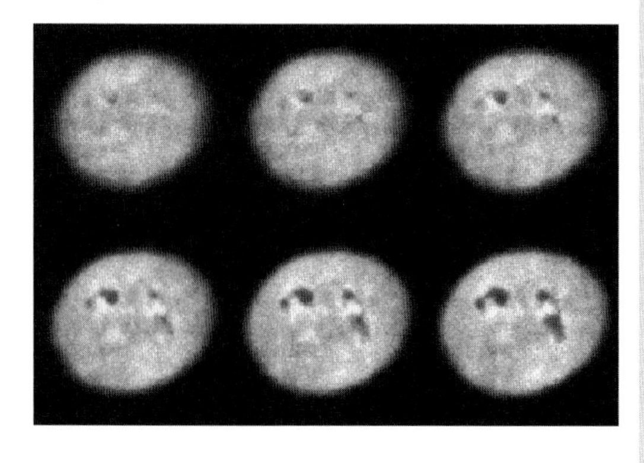

Figure 8.5 Nuclear data set (IP fluorescence). Six consecutive confocal sections through the nucleus of a *LAMA84* cell. The distance between each section is 0.2 µm, and the size of a pixel is 0.1 µm.

The confocal sections are collected every 0.2 µm, and the pixel size is 0.1 µm × 0.1 µm.

⇒ These sampling sizes are in agreement with Nyquist's criterion and guarantee that none of the details collected by the objective is lost.

8.2.2.2 AgNORs data set

The AgNOR proteins are imaged in laser scanning transmitted light microscopy. The image of the optical sections are collected simultaneously with the confocal sections of the PI fluorescence. The simultaneous acquisition of the nuclear data set and of the AgNORs data sets guarantees a good registration of the two sets.

⇒ The same laser source and the same scanning path are used for fluorescence and transmitted light. Only the detection path differs. Therefore, the confocal fluorescence sections and the transmitted light images are in perfect spatial correspondence.

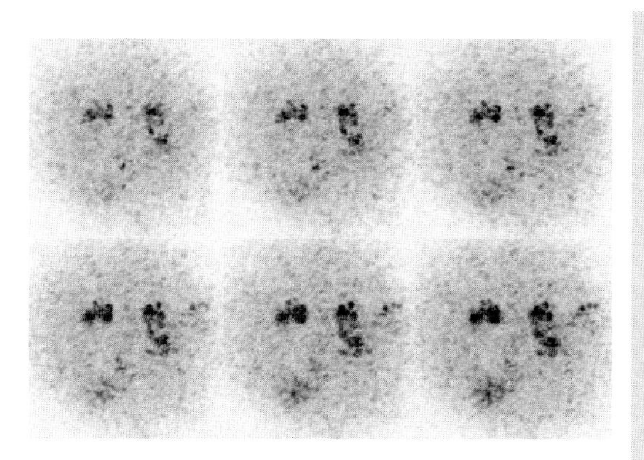

Figure 8.6 AgNOR data set (transmitted light). Six consecutive optical sections taken at the same level as those shown in Figure 8.5.

8.2.3 Segmentation of the Nucleus

8.2.3.1 Binary mask

The binary mask of the nucleus is obtained in a straightforward manner using a simple gray level threshold.

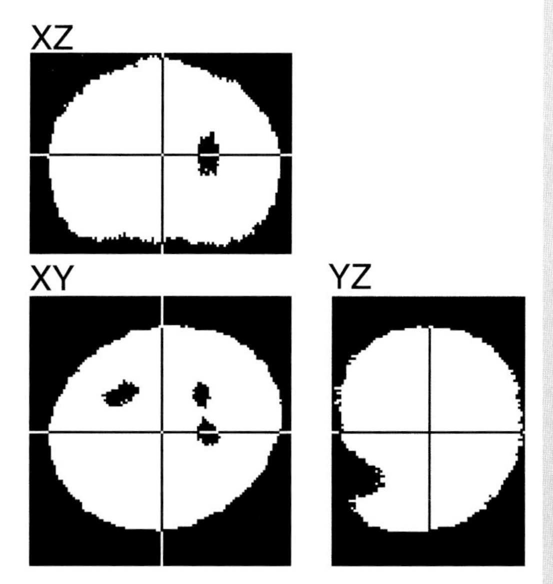

Figure 8.7 ***xy*, *xz*, and *yz* views of the raw segmentation mask of the nucleus. As can be seen on these sections, the binary mask of the nucleus displays unnatural cavities. These are imaging artifacts created by the silver aggregates, which create a shadow beneath them in the fluorescence volume. Therefore, this nuclear mask cannot be used for further image analysis.**

However, the obtained is far from being perfect. As we see in Figure 8.7, the basal part of the nucleus is mined by cavities created by the shadowing effect of AgNORs.

⇒ The AgNORs are silver aggregates; therefore, they are opaque to light. As they are in the way of the excitation light rays, they impair the fluorescence imaging DNA located immediately under them.

Fortunately, image processing helps us fill those cavities. This can be done either by iterative morphological closure (see Chapter 7) or by calculating the *convex hull* of the nuclear mask.

➪ The convex hull can be defined as the smallest convex polyhedron that contains the object. In plain English, it is similar to wrapping the nucleus in a plastic film. It sticks to the protrusions of the object and stretches above the cavities.

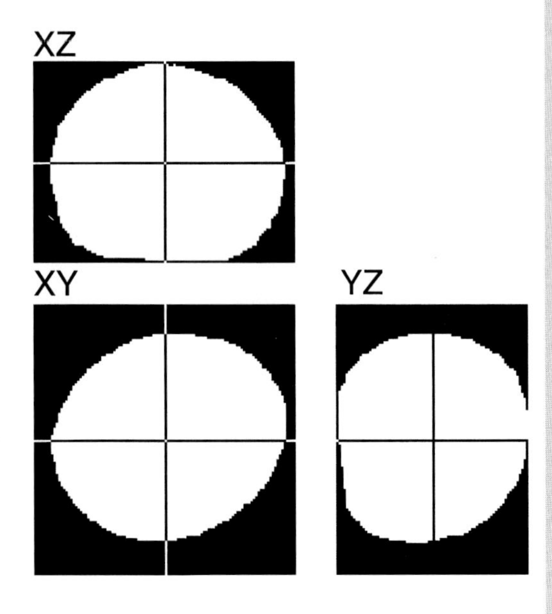

Figure 8.8 ***xy*, *xz*, and *yz* views of the segmentation mask of the nucleus after computation of the convex hull. The convex hull fills up the cavities and provides a binary mask with a regular shape.**

8.2.3.2 Definition of inner topology

In order to characterize the spatial distribution of silver aggregates in the nucleus, it is necessary to define a topological reference system. This is achieved by creating a map of the distance from the edge of the nucleus. The idea is to label most external layers of voxels of the nuclear mask with the distance value 1. (This means that the voxels of this layer are at distance 1 voxel from the background.) Then this layer is shrunk by the amount of 1 voxel. This gives a new layer labeled with the distance 2 and immediately inside layer 1. This shrinkage process is repeated until we reach the center of the nucleus.

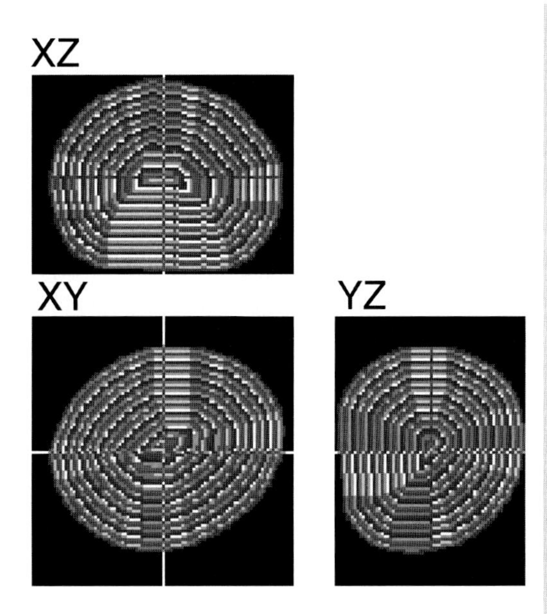

Figure 8.9 ***xy*, *xz*, and *yz* views of the "distance from edge" encoding of the convex hull mask. The isosurfaces are shown with a cyclic shade of gray as a function of the distance from the background.**

After this process, the nuclear mask is divided in nested layers organized like onion skins. Each skin or "distance isosurface" is labeled by its distance from the background.

8.2.4 Segmentation of AgNOR Aggregates

8.2.4.1 Resolution enhancement by digital deconvolution

One of the difficulties in this study is due to the difference of axial resolution provided by the two modes of microscopy. The nuclear data set is of high quality since the acquisition of fluorescence was done in confocal mode. Unfortunately, the AgNOR data are collected in simple transmitted light mode, which provides an appalling axial resolution. This low axial resolution makes it difficult to obtain a good segmentation of the silver aggregates. This limitation can be eliminated by digital deconvolution. As we described in Chapter 6, Section 6.2.4, it is possible to improve the axial resolution by reversing the process of image formation by the objective lens.

⇨ Unfortunately, a complete reversal of the imaging process is impossible, and what we obtain is a "fair estimate" of what the object is. Therefore, it is better to say that we achieve an improvement of the resolution rather than a "restoration."

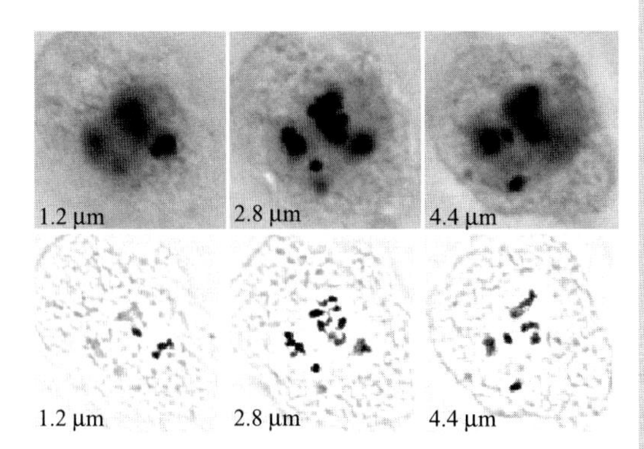

Figure 8.10 Comparison of raw transmitted light sections (top row) and deconvolved sections (bottom row). From left to right: optical sections at 1.2, 2.8, and 4.4 μm through the nucleus. Deconvolution removes blur and "out-of-focus" haze. The silver aggregates can be easily resolved and segmented.

Figure 8.10 illustrates the result of the digital deconvolution of the AgNOR data set. The improvement of resolution appears clearly. In the raw optical sections, the silver aggregates are hardly visible; they are drowning in "out-of-focus" haze. After deconvolution, these aggregates become clearly visible and are easily resolved.

8.2.4.2 Segmentation and labeling

After deconvolution, the AgNOR data sets are segmented using "simple gray-level thresholding algorithm (*see* Chapter 7). The value of the threshold is selected from the gray-level histogram.

⇨ The same threshold value is used for the segmentation of all cells in order.

The binary masks of the aggregates are labeled in order to extract them, one by one, and perform feature extraction. The location of the center of mass of each aggregate is stored for further topological analysis.

⇨ The center of mass is obtained as follows: the "object" voxels of an aggregate are characterized by their coordinates x_i, y_i, and z_i; thus, the coordinates of the center of mass G are

$$G_x = \frac{1}{n}\sum_{i=1}^{n} x_i,\ G_y = \frac{1}{n}\sum_{i=1}^{n} y_i,\ G_z = \frac{1}{n}\sum_{i=1}^{n} z_i,$$

where n is the number of "object" voxels.

8.2.5 Feature Extraction

8.2.5.1 Morphometric features

❑ *Nucleus*

- Volume

⇨ The volume is obtained by counting the number of voxels in the nuclear mask.

- Sphericity, which expresses how the nuclear shape diverges from an ideal sphere

⇨ This parameter is equal to the ratio of the radius of a virtual sphere of the same surface as the nuclear membrane vs. the radius of a virtual sphere of the same volume as the nucleus.

❑ AgNORs

- Number of aggregates
- Average volume of an aggregate and standard deviation
- Average volume fraction of an aggregate

⇨ It is the ratio of the average volume of an AgNOR aggregate vs. the volume of the nucleus.

- Total volume fraction of AgNORs

⇨ It is the ratio of the total volume of AgNORs vs. the volume of the nucleus; it expresses the relative amount of NOR proteins.

8.2.5.2 Topological features

Various topological parameters are derived from the positions of the aggregates, the center of mass of the nucleus, and the inner distance map. They all aim at characterizing how the aggregates are scattered inside the nucleus.

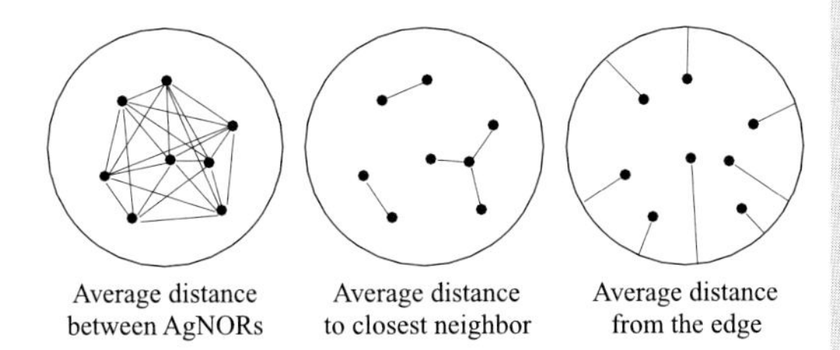

Figure 8.11 Some topological features.

Four main features are measured:

- The average distance between AgNORs, which expresses the clustering

⇨ The distance between each possible pair is measured and averaged.

- The average distance to the nearest aggregate, which expresses the compactness of AgNORs

⇨ For each aggregate, we find the closest neighbor. Therefore, it gives the average distance around a NOR protein where another NOR can be found.

- The angular grouping or anisotropy of AgNORs

⇨ This feature corresponds to the normalized modulus of the sum of the vectors which join the center of mass of the nucleus to aggregates.

- The average distance from the nuclear border (i.e., nuclear membrane)

8.2.6 Data Analysis

8.2.6.1 Factorial discriminant analysis

3-D image analysis provides large data sets. In the present study, 30 cell nuclei were analyzed per cell line and, for each nucleus, more than 10 features were measured. Multivariate data analysis tools are required to analyze such large data files. In our case, we used factorial discriminant analysis (FDA) to assess the description power of the different measured features. The graphical result of FDA is shown in Figure 8.12.

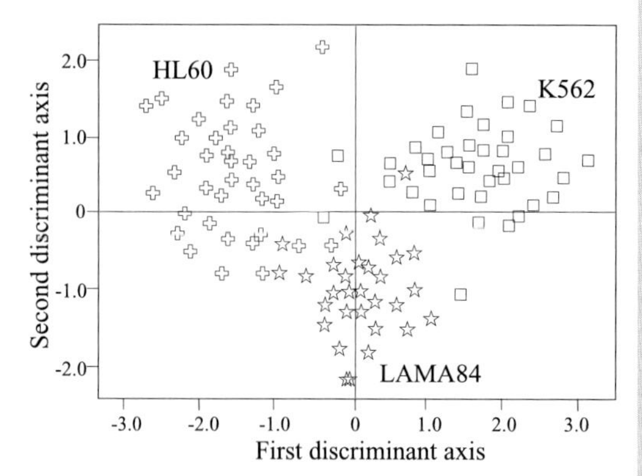

Figure 8.12 Factorial discriminant analysis. The measurements for each cell are projected in the factorial space defined by the two Fisher's linear discriminant axes. The cell populations are symbolized with different tokens:

☆ = LAMA84
+ = HL60
❑ = K562

In this graph, the horizontal and vertical axes are the Fisher's linear discriminant axes. These are linear combinations of the parameters that emphasize the differences between populations. We can see that the nuclei of the different cell lines form clearly separable clusters, meaning that the measured features are very representative of the differences in spatial organization of AgNORs in the different cell lines.

⇨ They define orthogonal directions within the feature space, along which the populations will appear separated as much as possible while each population appears homogeneous.

8.2.7 Material — Products/Solutions

8.2.7.1 Material

- Confocal laser scanning microscope

⇨ The emitted fluorescence is supplied with excitation (HeNe laser at 514 nm), dichroic (560 nm), and long-pass (590 nm) optical filters.

- Computer

⇨ Requires at least 14 Mbytes of RAM

- 3-D image analysis software

⇨ For image processing and feature extraction

8.2.7.2 Products/Solutions

- Cultures of leukematic cells

⇨ HL60, K562, and LAMA 84

- Culture medium (RPMI 1640, Boehringer)
- Fetal calf serum
- Paraformaldehyde (PFA)
- $AgNO_3$

⇨ Use deionized water.

- Thiosulfate
- Propidium iodide

⇨ Use a propidium iodide/RNase cocktail.

8.2.8 Protocol

8.2.8.1 Cell preparation

- Cells are cultured in supplemented culture medium.
  ⇨ Use culture medium supplemented with 10% fetal calf serum (FCS Gibco), 100 IU/mL penicillin, and 100 mg/mL streptomycin in humified atmosphere (5% CO_2, 95% air).
- Centrifuge cells. **5 min**
  ⇨ Centrifugation of 2 mL of suspension culture (1×10^6 cells) for 5 min at 800 rpm.
- Resuspend in PFA. **10 min**
  ⇨ Resuspend in 4% paraformaldehyde.
- Treat glass slides. **10 min**
  ⇨ The slides are treated with aminoalkysilane.
- Cytospin cells. **5 min**
  ⇨ Centrifuge at 700 rpm for 5 min using a Cytospin (Shandon). Approximately 5×10^4 cells are applied per slide.
- Wash the slides. **10 min**
  ⇨ Use deionized water.

8.2.8.2 NOR staining

- Cover the slides with fresh staining solution.
  ⇨ Mix equal volumes of the following solutions: (i) 0.5 mL/mL of $AgNO_3$ in deionized water; and (ii) 1 mg/mL of gelatin in 2% (by vol.) formic acid.
- Incubate the slides. **25 min at room temperature**
  ⇨ Avoid light exposure.
- Wash once in bidistilled water. **10 min**
- Rinse in thiosulfate. **10 min**
  ⇨ Use a 5% solution to stop the silver staining and act as a fixative.
- Rinse in water.

8.2.8.3 DNA staining

- Add propidium iodide. **30 min at 37°C**
  ⇨ 50 µg/mL fluorochrome
  ⇨ Avoid light exposure.
- Eliminate staining solution.
  ⇨ Avoid light exposure.
- Wash three times with distilled water. **5 min**
- Mount with antifading solution.
  ⇨ Use 2.3% w/v DABCO in 90% glycerol 0.1 *M* TRIS pH 7.5.

Index

Index

Q

R

S